Editorial Adviser: Nat Hiller

The
ELECTRICIAN'S
GUIDE

to the
17th Edition
of the
IEE Wiring Regulations
BS 7671: 2008
and Part P
of the
Building Regulations

**First Edition Updated
August 2008**

John Whitfield

Published by E·P·A Press
Wendens Ambo, Saffron Walden, Essex

Comments on earlier editions of the Electrician's Guide
- 'Whitfield's credentials are impressive, and for practical, on site use, the Electrician's Guide can be recommended'
- 'How to explain the inexplicable'
- 'Invaluable to students and electricians alike'
- 'Extremely user-friendly'

Additional copies of this book should be available through any good bookshop, electrical distributors and through the internet. In case of difficulty please contact the publishers directly at:

E·P·A Press
Bulse Grange
Wendens Ambo
Saffron Walden CB11 4JT
Tel: 01799 541207, Fax: 01799 541166
Email: enquiries@epapress.co.uk

Published by E·P·A Press, Wendens Ambo, Saffron Walden, UK

© John Whitfield 2008
Reprinted August 2008

EPA Press has published John Whitfield's Electrician's Guide to the IEE Wiring Regulations since 1991. The Guide has been regularly updated to reflect changes in the regulations. Hundreds of thousands of copies have been sold.

This Electrician's Guide, now to the 17th edition of the IEE Wiring Regulations, was first published in April 2008 and this August 2008 reprint is an update of the first April 2008 edition.

John Whitfield has many years' experience both as an electrical contractor and as a lecturer. He was for many years a senior lecturer and group leader at Norwich City College. He currently concentrates on writing and consultancy. He is a Chartered Engineeer of the IEE and member of CIBSE, and is the author of other bestselling guides.

Apart from fair dealing for the purposes of research or private study, or criticism or review, as permitted under the Copyright, Designs and Patents Act 1988, this publication may not be reproduced, stored or transmitted, in any form or by any means, without the prior written permission of the publisher.

Whilst the author and the publishers have made every effort to ensure that the information and guidance given in this book is correct, all parties must rely upon their own skill and judgment when making use of it. Neither the author nor the publishers assume any liability to anyone for any loss or damage caused by any error or omission in this book, whether such error or omission is the result of negligence or any other cause. Any and all such liability is disclaimed.

The author and the publishers are grateful to the Institution of Electrical Engineers for permission to reproduce extracts from its Regulations for Electrical Installations. These Regulations are definitive and should always be consulted in their original form in case of doubt.

British Library Cataloguing in Publication Data
Whitfield, John F. (Frederic), 1930 -
The electrician's guide to the 17th edition of the IEE wiring regulations BS: 7671 and to Part P of the Building Regulations
I. Title
1st Edition ISBN 978 0 9537885 5 2

Printed in the United Kingdom by Printwise

Bedfordshire County Council

9 39105396	
Askews	
621.31924	WHI

Contents

Acknowledgements

The author is indebted to the following for their help during the preparation of this Electrician's Guide:
Dr Katie Petty-Saphon, the Publisher, especially for her good-humoured forbearance.
Mr N Hiller, the Editor, for his helpful continued advice.
Mr D Harris, Chairman, of Kewtechnic Ltd for his help, particularly with Chapter 7.
Pamela Judge of Tektronix for supplying illustrations for Chapter 7.
Mr A W Cronshaw of the IET Technical Regulations Department for help and advice.
Mr John Aldridge of Aldridge Press for his help in publishing this edition.

Other publications available from E.P.A Press:
* The Guide to Electrical Maintenance
* The Guide to Electrical Safety at Work, 3rd Edition

Orders:
Copies of the above should be available through any good bookshop, electrical distributors and through the internet. In case of difficulty please contact the publishers directly at:
E.P.A Press
Bulse Grange
Wendens Ambo
Saffron Walden CB11 4JT
Fax: 01799 541166
Tel: 01799 541207
Email: enquiries@epapress.co.uk

Preface to the First Edition of the Electrician's Guide to the 17th Edition of BS 7671

When the 16th Edition of the IEE Regulations became BS 7671 it was widely assumed that there would never be a 17th Edition. This was proved wrong with the appearance of that publication in January 2008, and has led to my Guide to the 16th Edition being totally rewritten.

BS 7671 is an extremely comprehensive and lengthy document, and it would be impossible to deal with all of it in a publication of this size. I have attempted to interpret and explain all the Regulations that, in my opinion, will be of concern to the average electrician. Where doubt persists the original publication should be consulted. An aid to finding the required part of that publication will be the Regulation numbers provided in square brackets after the subheading to each section of this book.

My thanks are due, once again, to the many individuals and organisations that have made useful and constructive comments on previous editions of this book.

John Whitfield
Norwich
April 2008

CHAPTER ONE

The IEE Regulations, BS 7671 and this Guide

1.1 THE NEED FOR THIS ELECTRICIAN'S GUIDE

The BS 7671 Regulations published by the IET (IEE) summarise all requirements for electrical installations. The document has a legal character and is therefore not always easily understood and implemented by electricians.

The Author of this book has published a guide to the Regulations for more than 25 years. The level of detail has been carefully adjusted to ensure relevance as well as ease of understanding and implementation for practising electricians. Over the years, the Electrician's Guide to the Wiring Regulations has proved popular and successful.

This Guide to the 17th Edition is a continuation of the successful series. Again the content has been adjusted to ensure relevance and ease of understanding. It is intended to be accessible to all electricians, including the many electricians who are left to work on their own without the immediate backup of designers. For example, the formulas provided have been simplified in such a way that even electricians without advanced maths knowledge can perform the necessary calculations and make the correct installation decision. Students learning to become electricians will also benefit from this approach.

The Author of this Electrician's Guide realises that there is also the 'Electrical Installation Design Guide' published by the Institution of Electrical Engineers (IEE/IET) and BSI. Although the publication contains useful information, in the opinion of this Author, his Electrician's Guide offers a valuable alternative because of the assumptions and approach taken. That is, it is accessible to all electricians, including those working without the support of designers or without advanced maths knowledge, and to student electricians. Those accustomed to his previous editions will also find the style and format familiar.

This Electrician's Guide is not intended to replace the complete Regulations, although all the key regulations have been extracted and explained in the book. If an electrician wishes to master the subject, he/she should also equip him/herself with the Regulations and associated documents.

Note on Supply Voltage Level

For many years the supply voltage for single-phase supplies in the UK has been 240 V +/− 6%, giving a possible spread of voltage from 226 V to 254 V. For three-phase supplies the voltage was 415 V +/- 6%, the spread being from 390 V to 440 V. Most continental voltage levels have been 220/380 V.

In 1988 an agreement was reached that voltage levels across Europe should be unified at 230 V single phase and 400 V three-phase with effect from January 1st, 1995. In both cases the tolerance levels were then −6% to +10%, giving a single-phase voltage spread of 216 V to 253 V, with three-phase values between 376 V and 440V. On January 1st, 2008 the tolerance levels were widened to +/− 10%. This results in acceptable values from 207 V to 253 V for single-phase supplies, and 360 V to 440 V for three-phase supplies

Since the present supply voltages in the UK lie within the acceptable spread of values, Supply Companies in the UK have not reduced their voltages from 240/415 V. This is hardly surprising, because such action would immediately reduce the energy used by consumers (and the income of the Companies) by more than 8%.

Tables in the 17th Edition have, for the first time, been recalculated to take account of the unified voltage levels. Following this lead, calculations in this book have been based on supply voltages of 230 V and 400 V, but the electrician must bear in mind that true values are higher.

In due course, it is to be expected that manufacturers will supply appliances rated at 230 V for use in the UK. When they do so, there will be problems. A 230 V linear appliance used on a 240 V supply will take 4.3% more current and will consume almost 9% more energy. A 230 V rated 3 kW immersion heater, for example, will actually provide almost 3.27 kW when fed at 240 V. This means that the water will heat a little more quickly but there is unlikely to be a serious problem other than that the life of the heater may be reduced, the level of reduction being difficult to quantify. An appliance such as an immersion heater rated at 3 kW at 230 V but used on a 240 V supply will draw a current slightly in excess of 13 A. Long experience has shown us that 13 A plugs overheat when carrying 12.5 A, so here is yet another argument for not using plugs and sockets to feed 3 kW appliances.

Life reduction is easier to specify in the case of filament lamps. A 230 V rated lamp used at 240 V is likely achieve only 55% of its rated life (it will fail after about 550 hours instead of the average of 1,000 hours) but will be brighter and will run much hotter, possibly leading to overheating problems in some luminaires. The suggestion that filament lamps will be phased out on the grounds of inefficiency may well solve the problem, but this may be a long way off. The starting current for large concentrations of discharge lamps will increase dramatically, especially when they are very cold. High pressure sodium and metal halide lamps will show a significant change in colour output when run at higher voltage than their rating, and rechargeable batteries in 230 V rated emergency lighting luminaires will overheat and suffer drastic life reductions when fed at 240 V.

There could be electrical installation problems here for the future!

1.2 THE IEE REGULATIONS – BS 7671
1.2.1 International basis
All electricians are aware of 'The Regs'. For over a hundred years they have provided the rules that must be followed to ensure that electrical installations are safe. In 2007 the Institution of Electrical Engineers (the IEE) merged with the Institution of Electronics and Electrical Incorporated Engineers (IEEIE) to form the Institution of Engineering and Technology, known as the IET. However, the term 'IEE' is retained for the Regulations to prevent possible confusion.

A publication such as 'the Regs' must be regularly updated to take account of technical changes, and to allow for the 'internationalisation' of the Regulations. The ultimate aim is that all countries in the world will have the same wiring regulations. National differences make this still a dream, but we are moving slowly in that direction. The 15th Edition, when it was published in 1981, was the first edition of the IEE Regulations to follow IEC guidelines, and as such was novel in Great Britain. It was totally different from anything we had used before. The 16th and 17th Editions have not come as such a shock.

The international nature of the Regulations sometimes has strange results. For example, in {2.8} there are Regulations covering protection of installations from lightning – in fact, this work can be ignored in the UK because [443.2] points out that no action is necessary if the number of thunderstorm days for the region is less than 25 per annum. Since the UK (but not other parts of Europe) has less lightning activity than this level, these Regulations can be ignored here.

A word is necessary about the identification of parts of this Electrician's Guide and of BS 7671 – the IEE Regulations. In this Electrician's Guide, Regulation numbers are separated by a full point (full stop) and indicated by placing them in square brackets. Thus, [515.3.2] is the second Regulation in the third Section of Chapter 15. in Part 5. To avoid confusion, sections and sub-sections of this Electrician's Guide are divided by full points and enclosed in curly brackets. Hence, {5.4.6} is the sixth sub-section of section 4 of chapter 5 of this Electrician's Guide.

1.2.2 The Seventeenth Edition

The 17th Edition of BS 7671:2008 Requirements for Electrical Installations, also known as the 17th Edition of the IEE Wiring Regulations, was published on January 1st 2008, and all new electrical installations will be required to comply with them from July 1st, 2008.

The current trend is to move towards a set of wiring regulations with worldwide application. IEC publication 364 *Electrical Installations of Buildings* has been available for some time, and the 17th Edition is based on many of its parts. The European Committee for Electrotechnical Standardisation (CENELEC) uses a similar pattern to IEC 364 and to the Wiring Regulations, which, in the 17th Edition has moved still closer to it.

The introduction of the Free European Market in 1993 might well have caused serious problems for UK electrical contractors because, whilst the IEE Wiring Regulations were held in high esteem, they had no legal status that would require Europeans who were carrying out installation work in the UK to abide by them. This difficulty was resolved in October 1992 when the IEE Wiring Regulations became a British Standard, BS 7671, giving them the required international standing.

It does not follow that an agreed part of IEC 364 will automatically become part of the IEE Wiring Regulations. BS 7671 recognises all harmonised standards (or Harmonised Documents, HDs) that have been agreed by all member states of the European Union. BS EN standards are harmonised standards based on harmonised documents and are published without addition to or deletion from the original HDs. When a BS EN is published the relevant BS is superseded and is withdrawn. A harmonised standard, e.g. BS 7671, may have additions but not deletions from the original standard. IEC and CENELEC publications follow the pattern that will be shown in {1.2.3}, and it is not always easy to find which Regulations apply to a given application. For example, if we need to find the requirements for bonding, there is no set of Regulations with that title to which we can turn. Instead, we need to consider four separate parts of the Regulations, which in this case are:

1 [Chapter 13] Regulation [131.2.2]
2 [Section 415] Regulation [415.2]
3 [Section 514] Regulation [514.13]
4 [Sections 541, 542, 543 and 544] complete.

The question arises 'how do we know where to look for all these different Regulations'? The answer is two-fold. First, the Regulations themselves have a good index. Second, this Electrician's Guide also has a useful index, from which the applicable sub-section can be found. At the top of each sub-section in square brackets is the number(s) of the applicable Regulation(s).

The detail applying to a particular set of circumstances is thus spread in a number of parts of the Regulations, and the overall picture can only be appreciated after considering all these separate pieces of information. This Guide is particularly useful in drawing all this information together.

1.2.3 Changes due to the 17th Edition

Those who were conversant with the 16th Edition will find difficulty initially in navigating their way through the 17th Edition. This is because BS 7671 has been brought closer into alignment with CENELEC harmonisation documents, resulting in many of the Regulations being moved to new Parts and Chapters.

Some of the changes that will particularly affect the average electrician are listed below, but it must be stressed that this list is not comprehensive.

*   Requirements for safety services are included [35].
*   Protective measures to combat voltage disturbances and electromagnetic influences [Part 1].
*   The term 'direct contact' is now called 'basic protection' and 'indirect contact' becomes 'fault protection' [41].
*   Most socket outlet circuits are now required to be protected by 30 mA RCDs [41].
*   Tables of maximum earth-fault loop impedance have been modified assuming a supply voltage of 230 V rather than 240 V, giving slightly reduced values [Tables 41.2 to 41.4].
*   A new table giving maximum earth-fault loop impedances for RCD-protected circuits has been provided [Table 41.5].
*   Functional extra-low voltage (FELV) is now accepted as a protective measure [41].
*   Busbar trunking and powertrack systems are included for the first time [52].
*   30 mA RCD protection is now acceptable rather than metal covering for concealed cables outside accepted zones [52].
*   Metallic water pipes may now be used as earth electrodes in some cases [54].
*   Work on circuits with high protective currents has been moved from 'special installations' to the body of the Regulations [543.7].
*   A new set of Regulations covering lighting has been added [559].
*   The Regulations dealing with Safety Services have been extended [56].
*   Parts 6 (Special Installations) and 7 (Inspection and Testing) have changed places.
*   The minimum acceptable insulation resistance value has increased from 0.5 MΩ to 1.0 MΩ [612.3.2].
*   Requirements for low voltage generating sets are included [55].

- Expanded requirements for emergency escape lighting and fire protection are added [56].
- Highway power supplies have been moved from 'special installations' to the body of the Regulations [559].
- The zones for bathroom installations have been reclassified, each circuit in a bathroom now requires 30 mA RCD protection, and socket outlets can be installed in bathrooms provided that they are more than 3 m from the bath [701].
- The requirement for a shorter disconnection time for agricultural, horticultural and construction site installations intended to limit shock voltage to 25 V no longer applies and they now have a 50 V level like other installations [704 and 705].
- Socket outlets to feed caravan sites must now be individually RCD protected rather than in groups [708].
- Regulations for caravans have been moved to a new Section [708], away from caravan site installations [721].
- New Sections in Part 7 include:-
  Marinas and similar locations [709]
  Exhibitions, shows and stands [711]
  Solar voltaic (pv) power supply systems [712]
  Mobile or transportable units [717]
  Temporary installations for fairgrounds, amusement parks and circuses [740]
  Floor and ceiling heating systems [753]
- Eight new appendices are included, numbers 8 to 15 The titles will be found in {Table 1.2} earlier in this Chapter.

1.2.4 Plan of the Seventeenth Edition

The regulations are in seven parts as shown in {Table 1.1}.

Also included in the Regulations are 15 Appendices, listed in {Table 1.2}. The 17th Edition has a number of publications called 'Guides' which include much material previously to be found in Appendices. These Guides must be considered to form part of the Regulations; their titles are shown in {Table 1.3}.

Table 1.1 Arrangement of the 17th Edition Parts

Part 1	Scope, object and fundamental principles
Part 2	Definitions
Part 3	Assessment of general characteristics
Part 4	Protection for safety
Part 5	Selection and erection of equipment
Part 6	Inspection and testing
Part 7	Special installations or locations – particular requirements

Table 1.2 Appendices to the 17th Edition
The 17th Edition has eight more Appendices than the 16th.

Appendix 1	British Standards to which reference is made in The Regulations
Appendix 2	Statutory regulations and associated memoranda
Appendix 3	Time/current characteristics of overcurrent protective devices and residual current devices
Appendix 4	Current carrying capacity and voltage drop for cables and flexible cords
Appendix 5	Classification of external influences
Appendix 6	Model forms for certification and reporting

Appendix 7	Harmonised cable core colours
Appendix 8	Current carrying capacity and voltage drop for busbar trunking and powertrack systems
Appendix 9	Definitions – multiple source, d.c. and other systems
Appendix 10	Protection of conductors in parallel against overcurrent
Appendix 11	Effect of harmonic currents on balanced three-phase systems
Appendix 12	Voltage drop in consumers' installations
Appendix 13	Methods for measuring insulation resistance/impedance of floors and walls to earth or to the protective conductor system
Appendix 14	Measurement of fault loop impedance: consideration of the increase of the resistance of conductors with increase of temperature
Appendix 15	Ring and radial final circuit arrangements

The new Appendices 7 and 12 are of particular interest to electricians concerned with smaller installations.

Table 1.3 Guides and Guidance Notes to the Regulations
On-site guide
1   Selection and erection
2   Isolation and switching
3   Inspection and testing
4   Protection against fire
5   Protection against electric shock
6   Protection against overcurrents
7   Special installations and locations
8   Earthing and bonding

It is important to understand the relationship of the Appendices and of the Guidance Notes. Appendices provide information which the designer must have if his work is to comply with the Regulations. Other information, such as good practice, is contained in the Guidance Notes.

It is also important to have an understanding of the layout of the Regulations, so that work can be clearly identified. Each Part is divided into Chapters, which in turn are broken down into Sections within which are to be found the actual regulations themselves. The particular regulation is identified by a number, such as 612.13.2. Note that this Regulation number is spoken as 'six twelve point thirteen point two' and NOT as 'six hundred and twelve thirteen two'.

Note that whenever a group of digits is separated by a full point, the numbers represent a regulation; they are further identified in this Guide by placing them in square brackets, e.g. [612.13.2]. The apparent duplication of work within the Regulations may seem to be strange, but is necessary if the internationally agreed layout is to be followed.

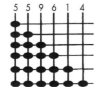

Part 5	Selection and erection of equipment
Chapter 55	Other equipment
Section 559	Luminaires and lighting installations
Group 559.6	Wiring systems
Regulation 559.6.1	Common rules
Regulation 559.6.1.4	Luminaire support couplers

The international nature of the Regulations sometimes provides some strange reading. For example, Regulations [411.6] deal with systems which are earthed using the IT method, described in {5.2.6} of this *Electrician's Guide*. However, such a system is not accepted for public supplies in the UK and is seldom used except in conjunction with private generators.

1.3 THE ELECTRICIAN'S GUIDE

1.3.1 The rationale for this Guide

In the first section of this chapter the reason for writing this Guide has been explained in general terms. It is important to understand that legally the whole of publication (BS 7671) must be followed; it is not permissible to extract one Regulation and to assume that if it is satisfied there is compliance with the whole of the Regulations. All electrical installation work must be carried out competently. The need for legal clarity does not always run parallel with technical clarity. The electrician, as well as his designer, if he has one, sometimes has difficulty in deciding exactly what a regulation requires, as well as the reasons for the regulation to be a requirement in the first place. The legal wording of the Regulations is paramount, and therefore it is not simplified to make its meaning more obvious to the electrician who is not expert in legal matters.

The object of this Electrician's Guide is to clarify the meaning of the Regulations and to explain the technical thinking behind them as far as is possible in a publication of this size. It is not possible to deal in detail with every Regulation. A choice has been made of those regulations which are considered to be most often met by the electrician in his everyday work. It must be appreciated that this Electrician's Guide cannot (and does not seek to) take the place of the Regulations, but merely to complement them

1.3.2 Using this Electrician's Guide

A complete and systematic study of this Electrician's Guide should prove to be of great value to those who seek a fuller understanding of the Regulations. However, it is appreciated that perhaps a majority of readers will have neither the time nor the inclination to read in such wide detail, and will want to use the book for reference purposes. Having found the general area of work, using either the contents list or the index of this Electrician's Guide, the text should help them to understand the subject. It has been previously stated that it is necessary for associated regulations to be spread out in various Parts and Chapters of the Wiring Regulations. In this *Electrician's Guide* they are collected together, so that, for example, all work on Earthing will be found in Chapter 5, on Testing and Inspection in Chapter 7, and so on. This *Electrician's Guide* is not a textbook of electrical principles. If difficulty is experienced in this respect, reference should be made to *Electrical Craft Principles* by the same author, published by the IET. This book has been written specifically for electricians.

CHAPTER TWO

Installation requirements and characteristics

2.1 INTRODUCTION

The first three parts of the Regulations lay the foundations for more detailed work later.

Part 1 **Scope, object and fundamental principles** applies to all installations and sets out their purposes as well as their status.

Part 2 **Definitions** details the precise meanings of terms used.

Part 3 **Assessment of general characteristics** is concerned with making sure that the installation will be fit for its intended purposes under all circumstances. The assessment must be made before the detailed installation design is started.

2.2 SAFETY REQUIREMENTS

2.2.1 Scope of the Regulations [110, Appendix 2]

Chapter 1 details exactly which installations are subject to the Regulations. They apply to all electrical installations except those listed later in this section. The 17th Edition includes a number of special installations not previously covered. These are

1 Marinas [8.18]
2 Mobile or transportable units [8.25]
3 Photovoltaic systems [8.22]
4 Low voltage generating sets [8.21]

Whilst the Regulations apply to them, some installations must also fulfil the requirements of appropriate British Standards. They are

1 High voltage luminous discharge tube installations (neon signs), BS 559 and BS EN 50107
2 Emergency lighting, BS 5266
3 Flameproof installations, BS EN 60079
4 Fire detection and alarm systems, BS 5839 (see {Chap 9})
5 Telephone and data cabling systems, BS 6701 Part 1
6 Surface heating systems BS 6351
7 Open cast mines and quarries, BS 6907

Some installations are excluded from the scope of the Regulations. These include

1 The supply and distribution by the Supply Companies
2 Railway traction, rolling stock and signalling equipment
3 Motor vehicles except for mobile units and caravans

4   Ship installations
5   Offshore platform installations, such as those for oil and gas extraction
6   Aircraft installations
7   Mine and quarry installations
8   Radio interference suppression equipment, except if it affects safety
9   Lightning protection systems to BS 6651
10   Parts of lift installations covered by BS 5655. These are concerned with the operation of the lift, not the supplies to it.
11   Electrical equipment of machines that are an internal part of their operation.

Where the electrical supply to an installation has been the subject of the Electrical Safety, Quality and Continuity Regulations 2002 (all Electrical Suppliers in England, Scotland and Wales are required to comply), it can be assumed that a permanent connection exists between the neutral and earth. In other cases, assurance must be obtained from the Supplier on this matter.

The Regulations apply to all installations fed at voltages included in the categories Extra-low-voltage and Low-voltage. The values of these classifications are shown in {Table 2.1}.

Table 2.1 Voltage classification
Extra-low-voltage – not exceeding 50 V ac or 120 V dc
Low-voltage – greater than extra-low voltage but not exceeding:
     1000 V ac or 1500 V dc between conductors
     600 V ac or 900 V dc from any conductor to earth

2.2.2 Legal status of the Regulations [120, Appendix 2]
The Regulations are intended to provide safety to people or to livestock from fire, shock and burns in any installation which complies with their requirements. They are not intended to take the place of a detailed specification, but may form part of a contract for an electrical installation. The Regulations themselves contain the legal requirements for electrical installations, whilst the Guidance Notes indicate good practice.

In premises licensed for public performances, such as theatres, cinemas, discos and so on, the requirements of the licensing authority will apply in addition to the Regulations. In mine and quarry installations the requirements of the Health and Safety Commission must be followed and are mandatory.

Domestic electrical installations have never previously been subject to the law in England and Wales, but they have been part of the Building Regulations in Scotland for some years. Part P has been added to the Building Regulations so that new and modified domestic installations must now comply with BS 7671 throughout England and Wales as well as Scotland. In practice this means that the Local Authority must be notified by the installers of the intention to carry out domestic electrical installation work unless they are members of a self-certification scheme. Some very minor work is excluded; full details will be found in {Chapter 11}.

The Electricity at Work Regulations 1989 became law on 1.1.92 in Northern Ireland and on 1.1.90 in the rest of the UK. Their original form made it clear that compliance with IEE Regulations was necessary, although it did not actually say so. It may be helpful to mention 'The Guide to Electrical Safety at Work' by John Whitfield, also published by EPA Press, which provides useful explanations of these Regulations.

The IEE Wiring Regulations became BS 7671 on 2nd October 1992 so that the legal enforcement of their requirements is easier, both in connection with the Electricity at Work Regulations and from an international point of view. The Construction (Design & Management) Regulations (CDM) were made under the Health & Safety at Work Act and implemented on 1.1.96. They require that installation owners and their designers consider health and safety requirements during the design and construction and throughout the life of an installation, including maintenance, repair and demolition. The Provision and Use of Work Equipment Regulations (1998), also made under the Health and Safety at Work Act, impose health and safety requirements on the use of machines, equipments, tools and installations which are used at work.

In 2002 the Electricity Safety, Quality and Continuity Regulations, usually called the ESQC Regulations, replaced the Electricity Supply Regulations. They affect BS 7671 in that an electrical installation certificate is likely to be required by the Electricity Supplier before a new installation is connected to his network. There are also implications where a private generator is to be used; whilst this option is not a common occurrence, it is likely to become so with the introduction of small gas powered generators, wind turbines, etc. (see {2.5} and {8.21}).

2.2.3 New inventions or methods [120.4]

The Regulations indicate clearly that it is intended that only systems and equipments covered by the Regulations should be used. However, in the event of new inventions or methods being applied, these should give at least the same degree of safety as older methods that comply directly.

Where such new equipments or methods are used, they should be noted on the Electrical Installation Certificate (see {7.8.2}) as not complying directly with the requirements of the Regulations.

2.2.4 Safety requirements [131, 52]

Safety is the basic reason for the existence of the Wiring Regulations. [131] has the title 'Fundamental Principles' and really contains, in shortened form, the full safety requirements for electrical installations. It has been said that the fourteen short regulations in [Section 131] are the Regulations and that the rest of the publication simply serves to spell out their requirements in greater detail. It is important to appreciate that the Electricity at Work Regulations apply to all electrical installations, covering designers, installers, inspectors, testers and users. The regular inspection and testing of all electrical installations is a requirement of the Electricity at Work Regulations.

Regulation [131.1] lists the faults in an electrical installation that are likely to lead to risk of injury to people or to livestock. They are

1 shock currents
2 excessive temperatures leading to burns, fires, etc.
3 explosion due to ignition of flammable gas
4 undervoltages, overvoltages and electromagnetic influences
5 unexpected movement of electrically operated equipment, such as sudden motor starting
6 interruptions of power supplies or of safety services
7 arcing causing eye damage, fire, high pressures or toxic gases

Perhaps the most basic rule of all, [134.1.1], is so important that it should be quoted more fully. It states:

'Good workmanship by competent persons and proper materials shall be used in the erection of the electrical installation.'

The details of [Chapter 13] are covered more fully later in this Guide.

The Building Regulations 1991 and the Building Standards (Scotland) Regulations 1990 require all new and refurbished dwellings to be fitted with mains operated smoke alarms. For a single family dwelling with no more than two floors there must be at least one alarm on each floor, installed within 7.5 m of the door to every habitable room. The alarms must have battery backup, must be interconnected so that detection of smoke by one operates all the alarms, and must be wired to a separate way in a distribution board or to a local, regularly used lighting circuit. Cables do not need special fire retardant properties and do not need to be segregated. The smoke alarm system must **not** be protected by an RCD. BS 5839 Part 6 "Code of practice for the design and installation of fire detectors in alarm systems" provides useful information, as does Appendix B of Guidance Note 4.

2.2.5 Design, Equipment, Installation and Testing [132, 133, 134]

Care must be taken to ensure that the design provides for the protection of people and of livestock from shock and from fire, and that it will function properly to meet the intended use. Proper information must be obtained to ensure that the available supply will be suitable and sufficient for the demands that will be placed on it. Special factors, such as exposure to bad weather conditions and possible explosion, must be taken into account in the design. The types and methods of completing the installation must be in accordance with these Regulations. When completed the installation must be inspected and tested to ensure that it complies with BS 7671. A recommendation concerning periodic inspection and testing must also be made. In effect, [Sections 132, 133 and 134]. are a summary of the complete Regulations.

Amendments to the Building Regulations have affected electrical installations in several ways. Firstly, Part P means that much electrical work in domestic installations is affected as described in detail in {Chapter 11}. Secondly, Part L affects the conservation of fuel and power. L1 covers dwellings, and requires that a 'reasonable' number of lighting outlets must be equipped with lamp holders or luminaires that will only accept lamps with a luminous efficacy exceeding 40 lumens per watt. This will rule out the use of filament lamps and encourage the installation of fluorescent (including compact fluorecent) lamps in homes. L2 covers offices, industrial and storage buildings, where the average lighting efficacy for the complete installation must exceed 40 1m/W. Due to their differing Building Regulations, other requirements may apply in Scotland.

2.2.6 Supplies for safety services [56]

Safety services are special installations which come into use in an emergency, to protect from, or to warn of, danger and to allow people to escape. Thus, such installations would include fire alarms , smoke and heat extraction, carbon monoxide warning and emergency lighting, supplies for sprinkler system pumps, as well as specially protected circuits to allow lifts to function in the event of fire. An electrical safety service is either

1 a non-automatic supply which is started by an operator, or
2 an automatic supply, which is independent of the operator.

Automatic supplies are classified in six categories depending on the time taken for the supply to become available. These vary from instant up to a break of 15 seconds.

Authorities other than the IEE Regulations will often require special needs for safety circuits, especially where people gather in large numbers. For example, safety circuits in cinemas are covered by the Cinematograph Regulations 1955, administered by the Home Office in England, Wales and Northern Ireland and by the Secretary of State in Scotland.

Safety circuits cannot be supplied by the normal installation, because it may fail in the dangerous circumstances the systems are there to guard against. The permitted sources of supply include cells and batteries, standby generators (see {2.5}) and separate feeders from the mains supply. The latter must only be used if it is certain that they will not fail at the same time as the main supply source.

The requirements of [Section 56] concerning safety services are extremely detailed and complex. Anyone concerned with the design or installation of such services is strongly advised to consult this Section of the Regulations.

The safety source must have adequate duration. This means, for example, that battery operated emergency lighting must stay on for the time specified in the applicable British Standard (BS 5266). Since such installations may be called on to operate during a fire, they must not be installed so that they pass through fire risk areas and must have fire protection of adequate duration. Safety circuits must be installed so that they are not affected by faults in normal systems and overload protection can be omitted to make the circuits less liable to failure. Safety sources must be in positions that are only open to skilled or instructed persons, and switchgear, control gear and alarms must be suitably labelled to make them clearly identifiable.

2.3 DEFINITIONS [Part 2, Appendix 9]

Any technical publication must make sure that its readers are in no doubt about exactly what it says. Thus, the meanings of the terms used must be defined to make them absolutely clear. For this purpose, [Part 2] includes over two hundred and fifty definitions of words used, as well as diagrams of the earthing systems. For example, the term 'low voltage' is often assumed to mean a safe level of potential difference. As far as the Regulations are concerned, a low voltage could be up to 1000 V ac or 1500 V dc!

Important definitions are those for skilled and instructed persons. A skilled person is one with technical knowledge or experience to enable the avoidance of dangers that may be associated with electrical energy dissipation. An instructed person is one who is adequately advised or supervised by skilled persons to enable dangers to be avoided. An ordinary person is one who is neither skilled nor instructed.

[Appendix 9], complete with numerous circuit diagrams, provides definitions for multiple source DC and other systems that are unlikely to be met by the average electrician.

The reader of the Regulations should always consult [Part 2] if in doubt, or suspects that there could possibly be a different meaning from the one assumed. If the word or phrase concerned is not included in [Part 2], BS 4727, 'Glossary of electrotechnical, power, telecommunication, electronics, lighting and colour terms' should be consulted.

2.4 ASSESSMENT OF GENERAL CHARACTERISTICS

2.4.1 Introduction [301.1]

Before planning and carrying out an electrical installation, the following characteristics must be considered.

1 the intended use, its structure and the supply available (see {2.4.2})
2 the external influences to which it will be exposed (see {2.4.3})
3 the compatibility of its equipment (see {2.4.4})
4 its maintainability (see {2.4.5})
5 the safety services required (see {2.2.6})
6 an assessment of the continuity of service

2.4.2 Purposes, supplies and structure [313, 314, 512]

[Section 313] includes regulations which effectively provide a check list for the designer to ensure that all the relevant factors have been taken into account.

An abbreviated check list would include the items shown in {Table 2.2}. Where equipments or systems have different types of supply, such as voltage or frequency, they must be separated so that one does not influence the other.

Table 2.2 Initial check list for the designer
a) the maximum demand
b) the number and type of live conductors
c) the type of earthing arrangement
d) the nature of supply, including:
   i) the nominal voltage(s)
   ii) the nature of current (a.c. or d.c.) and frequency
   iii) the prospective short circuit current at the supply intake
   iv) the earth fault loop impedance of the supply to the intake position
   v) the suitability of the installation for its required purposes, including maximum demand
   vi) type and rating of the main protective device (supply main fuse)
e) the supplies for safety services and standby purposes
f) the arrangement of installation circuits.

The requirements of the ESQC Regulations specify that the designer can require the electricity distributor to provide him with information regarding his proposed installation, as specified in {Table 2.2}, other than items a), d)v), e) and f). Correct assessment must be made of the required supplies for safety services, to ensure that they have adequate capacity, reliability and rating.

2.4.3 External influences [32, 51, Appendix 5]

This subject is a very clear example of the way that the IEE Wiring Regulations follow an international pattern. [Appendix 5] classifies external influences on an electrical installation using a code of two capital letters and a number. The first letter is one of three indicating the general category of external influence.

1 A Environment
2 B Utilisation
3 C Construction of buildings

The second letter refers to the nature of thee external influences, being different for each of the three categories as shown in {Table 2.3}. Next follows the number which indicates the class within each external influence. For example,

    BA5   B = utilisation
              A = capability
              5 = skilled persons

The ability of an enclosure to withstand the ingress of solid objects and of water is indicated by the index of protection (IP) system of classification. The system is detailed in BS EN 60529, and consists of the letters IP followed by two numbers. The first number indicates the degree of protection against solid objects, and the second against water. If, as is sometimes the case, either form of protection is not classified, the number is replaced with X. Thus, IPX5 indicates an enclosure whose protection against solid objects is not classified, but which will protect against water jets (see {Table 2.4}).

Table 2.3 External influences
(From BS 7671:2008, Appendix 5)

Environment (A)
AA   ambient temperature
AB   temperature and atmospheric humidity
AC   altitude
AD   presence of water
AE   foreign bodies
AF   corrosion
AG   impact
AH   vibration
AJ   other mechanical stresses
AK   flora (plants)
AL   fauna (animals)
AM   radiation
AN   solar (sunlight)
AP   seismic (earthquakes)
AQ   lightning
AR   wind

Utilisation (B)
BA   capability (such as physical handicap)
BB   resistance of human body
BC   contact with earth
BD   evacuation (such as difficult exit conditions)
BE   materials (fire risk)

Building (C)
CA   materials (combustible or non-flammable)
CB   structure (spread of fire etc.)

Other letters are also used as follows:
W   placed after IP indicates a specified degree of weather protection
S   after the numbers indicates that the enclosure has been tested against water penetration when not in use
M   after the numbers indicates that the enclosure has been tested against water penetration when in use.

Building utilization regarding evacuation is also covered.

BD1	Normal	Low density occupation, easy evacuation	Low height buildings used for habitation
BD2	Difficult	Low density occupation, difficult evacuation	High rise buildings
BD3	Crowded	High density occupation, easy evacuation	Theatres, cinemas, shops, etc.
BD4	Difficult & crowded	Low density occupation, difficult evacuation	High rise open to public, hotels, hospitals, etc.

There is also an impact protection code (the IK Code). These details are beyond the scope of this Guide but full details are found in Appendix B of the 2nd Ed. of Guidance Note 1. The EMC (Electromagnetic Compatibility) Regulations are now in force and require that electrical installations are designed and constructed so that they do not cause electromagnetic interference with other equipments or systems and are themselves immune to electromagnetic interference from other systems.

An abbreviated form of the information concerning the meanings of the two numbers of the IP system is shown in {Table 2.4}.

Table 2.4 Numbers in the I P system

First number	Mechanical protection against	Second number	Water protection against
0	Not protected	0	Not protected
1	Solid objects exceeding 50 mm	1	Dripping water
2	Solid objects exceeding 12 mm	2	Dripping water when tilted up to 15°
3	Solid objects exceeding 2.5 mm	3	Spraying water
4	Solid objects exceeding 1.0 mm	4	Splashing water
5	Dust protected	5	Water jets
6	Dust tight	6	Heavy seas
		7	Effects of immersion
		8	Submersion

One extremely important external influence which is often overlooked by the electrician (as well as the designer) is the effect of testing. The application of a potential of 500 V d.c. during an insulation test can be fatal to many parts of an electrical installation, such as movement indicators (PIRs), electronic starter switches, computers, etc. A sensible policy is to provide a clear and durable notice at the mains position listing the details and location of all pieces of equipment that must be disconnected before an insulation test is carried out.

2.4.4 Compatibility [331, 332, 512, 515]
One part of an electrical installation must not produce effects that are harmful to another part. For example, the heavy transient starting current of electric motors may result in large voltage reductions which can affect the operation of filament and discharge lamps. Again, the use of some types of controlled rectifier will introduce harmonics that may spread through the installation and upset the operation of devices such as electronic timers. Computers are likely to be affected by the line disturbances produced by welding equipment fed from the same system. 'Noisy' supplies, which contain irregular voltage patterns, can be produced by a number of equipments such as machines and thermostats. Such effects can result in the loss of data from computers, point-of-sale terminals, electronic office equipment, data transmission systems, and so on. Separate circuits may be necessary to prevent these problems from arising, together with the provision of 'clean earth' systems.

Very strict European laws limiting the amount of electromagnetic radiation permitted from electrical installations and appliances applied from January 1st, 1996 (the Electromagnetic Compatibility, or EMC, Directive. Equipment and installations must:

1 not generate excessive electromagnetic disturbances that could interfere with other equipments, and

2 have adequate immunity to electromagnetic disturbances to allow proper operation in its normal environment.

2.4.5 Maintainability [34, 529]

When designing and installing an electrical system it is important to assess how often maintenance will be carried out and how effective it will be. For example, a factory with a staff of fully trained and expertly managed electricians who carry out a system of planned maintenance may well be responsible for a different type of installation from that in a small motor-car repair works which does not employ a specialist electrician and would not consider calling one except in an emergency.

In the former case there may be no need to store spare cartridge fuses at each distribution board. The absence of such spares in the latter case may well lead to the dangerous misuse of the protective system as untrained personnel try to keep their electrical system working.

The electrical installation must be installed so that it can be easily and safely maintained, is always accessible for additional installation work, maintenance and operation and so that the built-in protective devices will always provide the expected degree of safety. Access to lighting systems is a common problem, and unless the luminaires can be reached from a stable and level surface, or from steps of a reasonable height, consideration should be given to the provision of hoisting equipment to enable the lighting system to be lowered, or the electrician raised safely to the luminaires. The Working at Height Regulations are important here.

For further information on the subject, see *The Guide to Electrical Maintenance*, by John Whitfield, published by E.P.A Press

2.5 LOW VOLTAGE GENERATING SETS [551, 552]

Where low voltage or extra-low voltage generating sets are used to power an installation, (a) as the sole means of supply, or (b) as a backup in case of failure of the supply, or (c) for use in parallel with the supply, these Regulations apply. Self-contained systems, operating at extra-low voltage, which include the source of energy (usually batteries) as well as the load are not covered. Small-scale embedded generators are becoming more common (see {8.21}).

Generating sets, for the purpose of these Regulations, include not only rotating machines powered by combustion engines, turbines and electric motors, but also photovoltaic cells (see {8.22}) (which convert energy from light into electricity) and electrochemical accumulators or batteries. Protection of circuits fed from generators must be no less effective than those applying to mains-fed systems. Voltage and frequency variations are much more likely with generators than with mains supplies, and it must be ensured that they do not cause danger or damage to the equipment.

Protection of persons and of equipment must be at least as effective in the case of an installation fed permanently or occasionally by a generating set as for a mains-fed installation. Special requirements for bonding apply where static invertors are used. A static invertor is an electronic system which produces an ac supply at a given voltage, frequency and waveform from a dc source (often from a battery). When two or more generating sets operate in parallel, circulating harmonic currents are a possibility. Such currents are at frequencies that are multiples of the normal supply frequency, and will possibly result in overloading of the connecting cables unless steps are taken to reduce or remove them. When not intended to run in parallel with the mains supply, for example when used as a standby system, interlocks and switching must be provided to ensure that parallel operation is not possible.

When a generating set is intended as a standby system for use in place of the mains supply in the event of failure, precautions must be taken to ensure that the generator cannot operate in parallel with the mains. Methods include an interlock between the operating systems of the changeover switches, a system of locks with a single transferable key, a three-position make-before-break changeover switch, or an automatic changeover switch with an interlock.

Where generators are intended to operate in parallel with the mains supply, the Supply Company must be consulted to ensure that the generator is compatible in all respects (including power factor, voltage changes, harmonic distortion, unbalance, starting, synchronising and voltage fluctuation) with the mains supply. In the event of any of the above parameters becoming incompatible with the mains supply, the generating set must be automatically disconnected. Such disconnection must also occur if its voltage or frequency strays outside the limits of protection. It must be possible to isolate the generating set from the mains supply, and that means of isolation must always be accessible to the supply company. See also {8.21}, small scale embedded generators.

2.6 STANDARDS [511, Appendix 1]

Every item of equipment that forms part of an electrical installation must be designed and manufactured so that it will be safe to use. To this end all equipment should, wherever possible, comply with the relevant British Standard; a list of the standards applying to electrical installations is included in the Regulations as [Appendix 1]. Increasingly we are meeting equipment that has been produced to the standards of another country. It is the responsibility of the designer to check that such a standard does not differ from the British Standard to such an extent that safety would be reduced. If the electrician is also the designer, and this may well happen in some cases, the electrician carries this responsibility.

Foreign standards may well be acceptable, but it is the responsibility of the designer to ensure that this is so. In the event of equipment not complying with the BS concerned, it is the responsibility of the installer to ensure that it provides the same degree of safety as that established by compliance with the Regulations.

A harmonised document (HD) is one agreed by all member nations of CENELEC. These documents may be published in individual countries with additions (BS 7671 is one such) but not with deletions. BS EN documents are Standards agreed by all CENELEC members and published without additions or deletions. Such EN standards supersede the original British Standard.

An increasing number of electrical contractors are registered as having satisfied the British Standards Institution that they comply fully with BS 5750 – Quality Assurance. This standard is not concerned directly with the Wiring Regulations, but since compliance with them is an important part of the organisation's assurance that its work is of the highest quality, registration under BS 5750 implies that the Wiring Regulations are followed totally.

2.7 UNDERVOLTAGE [445, 535]

These chapters of the Regulations deal with the prevention of dangers that could occur if voltage falls to a level too low for safe operation of plant and protective devices. Another problem covered is the danger that may arise when voltage is suddenly restored to a system that has previously been on a lower voltage or without voltage at all. For example, a machine that has stopped due to voltage falling to a low level may be dangerous if it restarts suddenly and unexpectedly when full voltage returns. A motor starter with built-in undervoltage protection will be explained in {8.16.1}.

The attention of the installer and the designer is drawn to the possibility that low voltage may cause equipment damage. Should such damage occur, it must not cause danger. Where equipment is capable of operating safely at low voltage for a short time, a time delay may be used to prevent switching off at once when undervoltage occurs. This system may prevent plant stoppages due to very short time voltage failures. However, such a delay must not prevent the immediate operation of protective systems.

2.8 VOLTAGE AND ELECTROMAGNETIC DISTURBANCES [44]

The 17th Edition has introduced a new [Chapter 44] with the title "Protection Against Voltage Disturbances and Electromagnetic Disturbances".

Voltages far in excess of those normally found within an installation can damage it. Such voltages, usually of very short duration and known as 'transients', are not uncommon, and can arise for three reasons.

1 Lightning strikes

Where the supply to an installation consists of external overhead wiring, or where such wiring forms a part of the installation, there is a danger that a lightning strike to it will occur. The resultant high voltage may then apply to parts of the installation, sometimes with disastrous consequences. High voltages can also appear within an installation as a result of a lightning strike close to it. BS 7671 does not mention the word 'lightning' in this amendment; it refers to 'overvoltages of atmospheric origin'. Direct lightning strikes to a supply network or to an electrical installation are not specifically mentioned.

2 Switching operations

When the supply to a load, particularly if it is highly inductive, such as a large electrical machine, or capacitive, such as power faction correction equipment, is suddenly switched off, the discharge of energy stored in the device may result in a very high voltage pulse of short duration. This is particularly likely where the supply is very suddenly switched, as is often the case with electronic controls.

3 Faults in high voltage and low voltage systems

There are four reasons for overvoltage as a result of faults in low voltage systems. They are

1 An earth fault in the supplying substation
2 Loss of the supply neutral
3 Short-circuit line to neutral in the low voltage installation
4 Accidental earthing of the line conductor in a low voltage IT system.

Where the supply to an installation is underground and there are no overhead lines within it, no special measures against lightning are necessary other than to ensure that the 'impulse withstand' of equipment is appropriate. A cable that is run overhead to an outbuilding, such as a catenary, is not classified as an overhead line for these purposes if it has insulated conductors and an earthed metal covering. If the installation includes an overhead line, or if the supply system includes such lines, protection must be provided if the installation is in an area with more than 25 thunderstorm days per year. In the UK our recorded thunderstorm days are less than 25 per year, so no special measures are required. This is certainly not the case for other parts of Europe. Alternatively, the decision concerning protection may be based on a risk assessment method, a statistical process of determining probable future events based on past events.

The method of protection, against lightning and switching transients and faults, is to fit a surge protective device, which will detect and prevent voltage surges in excess of 1.5 kV. This device is effectively a voltage dependent resistor (VDR) connected across phase(s) and neutral as close as possible to the origin of the installation. Such a device has a very high resistance at normal voltage levels so that the current it carries is negligible. When it 'sees' an excessive voltage, its resistance reduces sharply so that it effectively short-circuits the system for the very short duration of the overvoltage event.

CHAPTER THREE

Installation control and protection

3.1 INTRODUCTION [410, 421, 430]

Electrical installations must be protected from the effects of short circuit and overload. In addition, the people using the installations, as well as the buildings containing them, must be protected from the effects of electric shock, fire and of other hazards arising from faults or from misuse.

Not only must automatic fault protection of this kind be provided, but an installation must also have switching and isolation which can be used to control it in normal operation, in the event of emergency, and when maintenance is necessary. This Chapter will consider those Regulations that deal with the disconnection of circuits, by both manual and automatic means, the latter in the event of shock, short circuit or overload. It does not include the Regulations that concern automatic disconnection in the event of an earth fault: these are considered in {Chapter 5}.

In order that anyone operating or testing the installation has full information concerning it, a diagram or chart must be provided at the mains position showing the number of points and the size and type of cables for each circuit, the method of providing protection from normal conditions, (which was previously known as direct contact in the 16th Edition) (see {3.4.5}) and details of any circuit in which there is equipment, such as passive infra-red detectors, electronic fluorescent starters, or other devices that are vulnerable to the high voltage used for insulation testing.

3.2 SWITCHING

3.2.1 Switch positions [131, 530, 536 and 537]

A switch is defined as a device which is capable of making, carrying and breaking a circuit under normal and under overload conditions. It can make, but will not necessarily break, a short circuit, which should be broken by the overload protecting fuse or circuit breaker. A switching device may be marked with ON and OFF positions, or increasingly, the numbers 1 for ON and 0 for OFF are being used.

A semiconductor device is often used for switching some lighting and heating circuits, but will not be suitable for disconnecting overloads; thus, it must be backed up by a mechanical switch. The semiconductor is a functional switch but must NOT be used as an isolator.

{Figure 3.1} shows which poles of the supply need to be broken by the controlling switches. For the TN-S system (earth terminal provided by the Electricity Company), the TN-C-S system (protective multiple earthing) and the TT system (no earth provided at the supply), all phase conductors MUST be switched, but NOT the protective (earth) conductor.

The neutral conductor need not be broken except for:
1 the main switch in a single-phase installation, *or*
2 heating appliances where the element can be touched, *or*
3 autotransformers (not exceeding 1.5 kV) feeding discharge lamps

The neutral will need to be disconnected for periodic testing, and provision must be made for this; it is important that the means of disconnection is accessible and can only be completed with the use of a tool. The protective conductor should never be switched, except when the supply can be taken from either of two sources with earth systems which must not be connected together. In this case the switches needed in the protective conductors must be linked to the phase switches so that it is impossible for the supply to be provided unless the earthing connection is present.

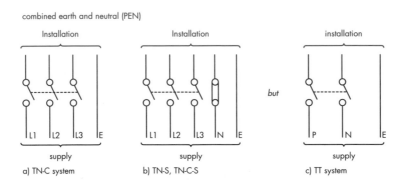

Fig 3.1 Supply system broken by switches
(a) TN-C system (b) TN-S, TN-C-S systems (c) TT systems

Every circuit must be provided with a switching system so that it can be interrupted on load. In practice, this does not mean a switch controlling each separate circuit; provided that loads are controlled by switches, a number of circuits may be under the overall control of one main switch. An example is the consumer unit used in the typical house, where there is usually only one main switch to control all the circuits, which are provided with individual switches to operate separate lights, heaters, and so on. If an installation is supplied from more than one source there must be a separate main switch for each source, and each must be clearly marked to warn the person switching off the supplies that more than one switch needs to be operated.

It should be noted that a residual current device (RCD) may be used as a switch provided that its rated breaking current is high enough. It is of the utmost importance that all switches and isolators are clearly identified to indicate their functions. [Table 53], which is new to the 17th Edition, gives extensive guidance on the selection of protective, isolation and switching devices.

3.2.2 Emergency switching [537.4, 537.6]
Emergency switching is defined as rapidly cutting the supply to remove hazards. For example, if someone is in the process of receiving an electric shock, the first action of a rescuer should be to remove the supply by operating the emergency switch, which may well be the main switch. Such switching must be available for all instal-

lations. Note that if there is more than one source of supply a number of main switches may need to be opened (see {3.2.1}). The designer must identify all possible dangers, including electric shock, mechanical movement, excessive heat or cold and radiation dangers, such as those from lasers or X-rays. Means of operation (handles, pushbuttons etc.) should be clearly identified, preferably by using the colour red.

In the special case of electric motors, the emergency switching must be adjacent to the motor. In practice, such switching may take the form of a starter fitted close to the motor, or an adjacent stop button (within 2 m) where the starter is remote. Where a starter or contactor is used as an emergency switch, a positive means must be employed to make sure that the installation is safe. For example, operation should be when the operating coil is de-energised, so that an open circuit in the coil or in its operating circuit will cause the system to be switched off {Fig 3.2}. This is often called the 'fail-safe' system.

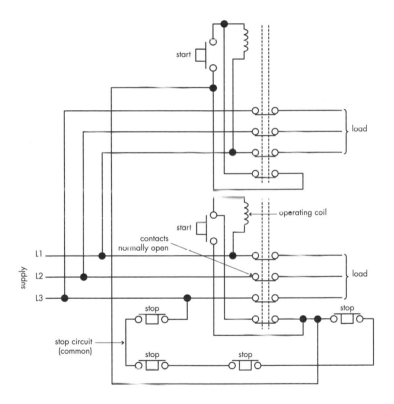

Fig 3.2 Two circuit breakers linked to a common stop circuit. The system is 'fail-safe'.

To prevent unexpected restarting of rotating machines, the 'latching off' stop button shown in {Fig 3.3} is sometimes used. On operation, the button locks (latches) in the off position until a positive action is taken to release it.

In single-phase systems, it must be remembered that the neutral is earthed. This means that if the stop buttons are connected directly to the neutral, a single earth fault on the stop button circuit would leave the operating coil permanently fed and prevent the safety system from being effective. It is thus essential for the operating coil to be directly connected to the neutral, and the stop buttons to the phase. Such an earth fault would then operate the protective device and make the system safe.

The means of emergency switching must be such that a single direct action is required to operate it. The switch must be readily accessible and clearly marked in a way that will be durable. Consideration must be given to the intended use of the premises in which the switch is installed to make sure as far as possible that the switching system is always easy to reach and to use. For example, the switch should not be situated at the back of a cupboard which, in use, is likely to be filled with materials making it impossible to reach the switch.

In cases where operation could cause danger to other people (an example is where lighting is switched off by operating the emergency switch), the switch must be available only for operation by skilled or instructed persons. Every fixed or stationary appliance must be provided with a means of switching which can be used in an emergency. If the device is supplied by an unswitched plug and socket, withdrawal of the plug is NOT acceptable to comply with this requirement; such action is acceptable for functional switching {3.2.4}.

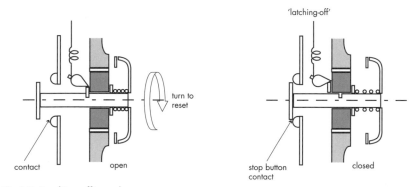

Fig 3.3 'Latching-off' stop button

Where any circuit operates at a p.d. (potential difference) exceeding low voltage a fireman's switch must be provided. Such installations usually take the form of discharge lighting (neon signs), and this requirement applies for all external systems as well as internal signs that operate unattended. The purpose is to ensure the safety of fire fighters who may, if a higher voltage system is still energised, receive dangerous shocks when they play a water jet onto it. The fireman's switch is not required for portable signs consuming 100 W or less which are supplied via an easily accessible plug and socket.

The fireman's switch must meet the following requirements

1   The switch must be mounted in a conspicuous position not more than 2.75m from the ground.

2   It must be coloured red and have a label at least 150 mm x 100 mm in 36 point lettering reading 'FIREMAN'S SWITCH'. On and off positions should be clearly marked, and the OFF position should be at the top. A lock or catch should be provided to prevent accidental reclosure.

3   For exterior installations the switch should be near the load, or to a notice in such a position to indicate clearly the position of the well-identified switch.

4   For interior installations, the switch should be at the main entrance to the building.

5 Ideally, no more than one internal and one external switch must be provided. Where more become necessary, each switch must be clearly marked to indicate exactly which parts of the installation it controls.

6 Where the local fire authority has additional requirements, these must be followed.

7 The switch should be arranged on the supply side of the step-up sign transformer.

3.2.3 Switching for mechanical maintenance [537.3]

Mechanical maintenance is taken as meaning the replacement and repair of non-electrical parts of equipment, as well as the cleaning and replacement of lamps. Thus, we are considering the means of making safe electrical equipment which is to be worked on by non-electrical people.

Such switches must be:

1 easily and clearly identified by those who need to use them

2 arranged so that there can be no doubt when they are on or off

3 near the circuits or equipments they switch

4 able to switch off full load current for the circuit concerned

5 arranged so that it is impossible for them to be reclosed unintentionally.

Where mechanical maintenance requires access to the interior of equipment where live parts could be exposed, special means of isolation are essential. Lamps should be switched off before replacement.

3.2.4 Functional switching [537.5]

Functional switching is used in the normal operation of circuits to switch them on and off. A switch must be provided for each part of a circuit that may need to be controlled separately from other parts of the installation. Such switches are needed for equipment operation, and a group of circuits can sometimes be under the control of a single switch unless separate switches are required for reasons of safety. Semiconductor switches may control the current without opening the poles of the supply. Off-load isolators, fuses and links must not be used as a method of switching. A plug and socket may be used as a functional switching device provided its rating does not exceed 16 A. It will be appreciated that such a device must not be used for emergency switching (see {3.2.2}).

3.3 ISOLATION

3.3.1 Isolator definition [537.2]

An isolator is not the same as a switch. It should only be opened when not carrying current, and has the purpose of ensuring that a circuit cannot become live whilst it is out of service for maintenance or cleaning. The isolator must break all live supply conductors; thus both phase and neutral conductors must be isolated. It must, however, be remembered that switching off for mechanical maintenance (see {3.2.3}) is likely to be carried out by non-electrically skilled persons and that they may therefore unwisely use isolators as on-load switches. To prevent an isolator, which is part of a circuit where a circuit breaker is used for switching, from being used to break load current, it must be interlocked to ensure operation only after the circuit breaker is already open. In many cases an isolator can be used to make safe a particular piece of apparatus whilst those around it are still operating normally.

3.3.2 Isolator situation [537.2]

Isolators are required for all circuits and for all equipments, and must be adjacent to the system protected.

As for a switch, the isolator must be arranged so that it will not close unintentionally due to vibration or other mechanical effect. It must also, where remote from the circuit protected, be provided with a means of locking in the OFF position to ensure that it is not reclosed whilst the circuit is being worked on. The isolator should also be provided, where necessary, with means to prevent reclosure in these circumstances. This may be a padlock system to enable the device to be locked in the OFF position, a handle which is removable in the off position only, etc. It is particularly important for motors and their starters to be provided with adjacent isolation to enable safe maintenance.

The special requirements for isolating discharge lamp circuits operating at voltages higher than low voltage will be considered in {8.13.2}.

3.3.3 Isolator positions [537.2]

Every circuit must have its means of isolation, which must be lockable in the OFF position where remote from the equipment protected. The OFF position must be clearly marked in all cases so that there is no doubt in the mind of the operator as to whether his circuit is isolated and thus safe to work on. Adequate notices and labels must be displayed to ensure safe and proper isolation and to supplement locks to prevent inadvertent re-closing of isolators when this could cause danger. For single-phase systems, both live conductors (phase and neutral) must be broken by the isolator (TT system). On three phase supplies, only the three phases (L1, L2 and L3) need to be broken, the neutral being left solidly connected (TN-S and TN-C-S systems).

This neutral connection is usually through a link that can be removed for testing. Clearly, it is of great importance that the link is not removed during normal operation, so it must comply with one or both of the following requirements:

1 it can only be removed by using tools, and/or

2 it is accessible only to skilled persons.

In many circuits capacitors are connected across the load. There are many reasons for this, the most usual in industrial circuits being power factor correction. When the supply to such a circuit is switched off, the capacitor will often remain charged for a significant period, so that the isolated circuit may be able to deliver a severe shock to anyone touching it. The Regulations require that a means of discharging such capacitors should be provided. This usually takes the form of discharge resistors {Fig 3.4}, which provide a path for discharge current. These resistors are connected directly across the supply, so give rise to a leakage current between live conductors. This current is reduced by using larger resistance values, but this increases the time taken for the capacitors to discharge to a safe potential difference. In practice, a happy medium is struck between these conflicting requirements, resistors with values of about 100 kΩ being common.

When isolating a circuit so that it can be worked on, care must be taken to ensure that there is no uninterruptible power supply (UPS) or other standby system which may make it dangerous although isolated from the mains supply.

3.3.4 Semiconductor isolators [537.2]

Semiconductors are very widely used in electrical installations, from simple

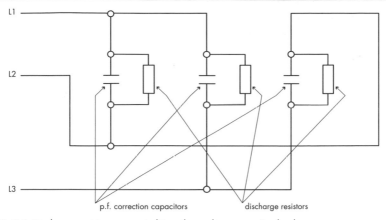

Fig 3.4 Discharge resistors connected to a three-phase capacitor bank

domestic light dimmers (usually using triacs) to complex speed controllers for three phase motors (using thyristors). Whilst the semiconductors themselves are functional switches, operating very rapidly to control the circuit voltage, they must NOT be used as isolators.

This is because when not conducting (in the OFF position) they still allow a very small leakage current to flow, and have not totally isolated the circuit they control. {Figure 3.5} shows semiconductors in a typical speed control circuit.

3.3.5 Isolator identification [537.2]

The OFF position on all isolators must be clearly marked and should not be indicated until the contacts have opened to their full extent to give reliable isolation.

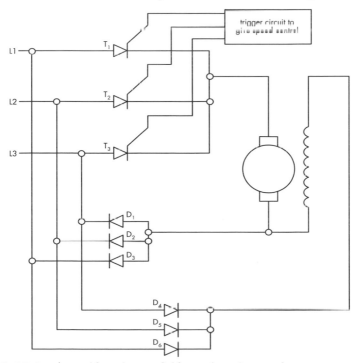

Fig 3.5 Speed control for a dc motor fed from a three-phase supply

Every isolator must be clearly and durably marked to indicate the circuit or equipment it protects. If a single isolator will not cut off the supply from internal parts of an enclosure, it must be labelled to draw attention to the possible danger. Where the unit concerned is suitable only for off-load isolation, this should be clearly indicated by marking the isolator 'Do NOT open under load'.

3.4 ELECTRIC SHOCK PROTECTION

3.4.1 The nature of electric shock

The nervous system of the human body controls all its movements, both conscious and subconscious. The system carries electrical signals between the brain and the muscles, which are thus stimulated into action. The signals are electro-chemical in nature, with levels of a few millivolts, so when the human body becomes part of a much more powerful external circuit, its normal operations are swamped by the outside signals. The current forced through the nervous system of the body by external voltage is electric shock.

All the muscles affected receive much stronger signals than those they normally get and operate very much more violently as a result. This causes uncontrolled movements and pain. Even a patient who is still conscious is usually quite unable to counter the effects of the shock, because the signals from his brain, which try to offset the effects of the shock currents, are lost in the strength of the imposed signals. A good example is the 'no-let-go' effect. Here, a person touches a conductor which sends shock currents through his hand. The muscles respond by closing the fingers on the conductor, so it is tightly grasped. The person wants to release the conductor, which is causing pain, but the electrical signals from his brain are swamped by the shock current, and he is unable to let go of the offending conductor.

The effects of an electric shock vary considerably depending on the current imposed on the nervous system, and the path taken through the body. The subject is very complex but it has become clear that the damage done to the human body depends on two factors:

1 the value of shock current flowing, and
2 the time for which it flows.

These two factors and the production of the IEC publication 61140 have governed the international movement towards making electrical installations safer.

3.4.2 Resistance of the shock path [410]

In simple terms the human body can be considered as a circuit through which an applied potential difference will drive a current. As we know from Ohm's Law, the current flowing will depend on the voltage applied and the resistance of the current path. Of course, we should try to prevent or to limit shock by aiming to stop a dangerous potential difference from being applied across the body. However, we have to accept that there are times when this is impossible, so the important factor becomes the resistance of the current path.

The human body is composed largely of water, and has very low resistance. The skin, however, has very high resistance, the value depending on its nature, on the possible presence of water, and on whether it has become burned. Thus, most of the resistance to the passage of current through the human body is at the points of entry and exit through the skin. A person with naturally hard and dry skin will offer much higher resistance to shock current than one with soft and moist skin; the skin resistance becomes very low if it has been burned, because of the presence of conducting particles of carbon.

In fact, the current is limited by the impedance of the human body, which includes self-capacitance as well as resistance. The impedance values are very difficult to predict, since they depend on a variety of factors including applied voltage, current level and duration, the area of contact with the live system, the pressure of the contact, the condition of the skin, the ambient and the body temperatures, and so on.

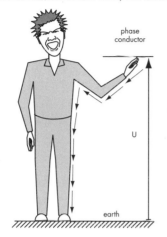
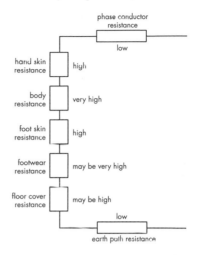

Fig 3.6 Path of electric shock current

Figure 3.6 is a simplified representation of the shock path through the body, with an equivalent circuit which indicates the components of the resistance concerned. It must be appreciated that the diagram is very approximate; the flow of current through the body will, for example, cause the victim to sweat, reducing the resistance of the skin very quickly after the shock commences. Fortunately, people using electrical installations rarely have bare feet, and so the resistance of the footwear, as well as of the floor coverings, will often increase overall shock path resistance and reduce shock current to a safer level.

Guidance Note 7 (Special Locations) provides data on the impedance of the human body. However, the figures are complicated by the fact that values differ significantly from person to person; it would be sensible to assume a worst case possibility which suggests that the impedance of the human body from hand to foot is as low as 500 Ω. Since this calculates to a body current of 460 mA when the body has 230 V applied, we are considering a fatal shock situation.

There are few reliable figures for shock current effects, because they differ from person to person, and for a particular person, with time. However, we know that something over one milliampere of current in the body produces the sensation of shock, and that one hundred milliamperes is likely quickly to prove fatal, particularly if it passes through the heart.

If a shock persists, its effects are likely to prove to be more dangerous. For example, a shock current of 500 mA may have no lasting ill effects if its duration is less than 20 ms, but 50 mA for 10 s could well prove to be fatal. The effects of the shock will vary, but the most dangerous results are ventricular fibrillation (where the heart beat sequence is disrupted) and compression of the chest, resulting in a failure to breathe. The resistance of the shock path is of crucial importance. The Regulations insist on special measures where shock hazard is increased by a reduction in body resistance and good contact of the body with earth potential. Such situations include locations containing bathtubs or showers, swimming pools, saunas and so on. The Regulations applying to these special installations are considered in {Chapter 7 Inspection and Testing}.

Another important factor to limit the severity of electric shock is the limitation of earth fault loop impedance. Whilst this impedance adds to that of the body to reduce shock current, the real purpose of the requirement is to allow enough current to flow to operate the protective device (RCD) and thus to cut off the shock current altogether quickly enough to prevent death from shock.

How quickly this must take place depends on the level of body resistance expected. Where sockets are concerned, the portable appliances fed by them are likely to be grasped firmly by the user so that the contact resistance is lower. Thus, disconnection within 0.4 s is required. In the case of circuits feeding fixed equipment, where contact resistance is likely to be higher, the supply must be removed within 5 s. Earth fault loop impedance is considered more fully in {Chapter 5}.

3.4.3 Contact with live conductors [410]

In order for someone to get an electric shock he or she must come into contact with a live conductor. Two types of contact are classified.

1 Basic protection (formerly known as direct contact)

An electric shock results from contact with a conductor which forms part of a circuit and would be expected to be live. A typical example would be if someone removed the plate from a switch and touched the phase (live) conductors inside (see {Fig 3.7}). Overcurrent protective systems will offer no protection in this case, but it is possible that an RCD with an operating current of 30 mA or less may do so.

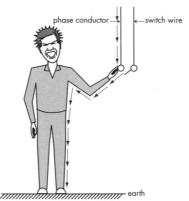

Fig 3.7 Basic protection (was direct contact)

2 Protection under fault conditions (formerly known as indirect contact).

An electric shock is received from contact with something connected with the electrical installation that would not normally be expected to be live, but has become so as the result of a fault. This would be termed an exposed conductive part. Alternatively, a shock may be received from a conducting part which is totally unconnected with the electrical installation, but which has become live as the result of a fault. Such a part would be called an extraneous conductive part.

An example illustrating both types of contact under fault conditions is shown in {Fig 3.8}. Danger in this situation results from the presence of a phase to earth fault on the kettle. This makes the kettle case live, so that contact with it, and with a good earth (in this case the tap) makes the human body part of the shock circuit.

The severity of the shock will depend on the effectiveness of the kettle protective conductor system. If the protective system had zero resistance, a 'dead short' would be caused by the fault and the protecting fuse or circuit breaker would open the circuit. The equivalent circuit shown in {Fig 3.8(b)} assumes that the protective conductor has a resistance of twice that of the phase conductor at 0.6 Ω, and will result in a potential difference of 153 V across the victim. The higher the protective circuit resistance, the greater will be the shock voltage, until an open circuit protective system will result in a 230 V shock.

If the protective conductor had zero resistance during the short time it took for the circuit to open, the victim would be connected across a zero resistance which would result in no volt drop regardless of the level reached by the fault current, so there could be no shock.

The shock level thus depends entirely on the resistance of the protective system. The lower it can be made, the less severe will be the shocks that may be received.

To sum up this subsection: Basic protection contact is contact with a live system which should be known to be dangerous and contact under fault conditions concerns contact with metalwork which would be expected to be at earth potential, and thus safe. The presence of socket outlets close to sinks and taps is not prohibited by the IEE Wiring Regulations, but could cause danger in some circumstances. It is suggested that special care be taken, including consultation with the Health and Safety Executive in industrial and commercial situations.

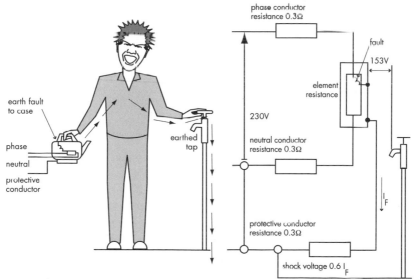

Fig 3.8 Fault protection contact:
(Left-side) fault condition (Right-side) equivalent circuit, assuming a fault resistance of zero

3.4.4 Protection from contact [41]

Warning notices that draw attention to the danger of electric shock are required:

a) Inside an item of equipment or enclosure where dangerous voltage is present
b) Between simultaneously accessible items of equipment or enclosures
c) Where different nominal voltages are accessible.

Four methods of protection are listed in the Regulations.

1 Protection by separated extra-low voltage (SELV)

This voltage is electrically separated from earth and from other systems, is provided by a safety source, and is low enough (not exceeding 50 V ac or 120 V dc) to ensure that contact with it cannot produce a dangerous shock in people with normal body resistance or in livestock. The system is uncommon. See also section {8 .17.2}.

2 Protection by protective extra-low voltage (PELV)

The method has the same requirements as SELV but is earthed at one point. Protection against contact may not be required if the equipment is in a building, if the output voltage level does not exceed 25 V rms ac or 60 V basic ripple-free dc in normally dry locations, or 6 V rms ac or 15 V ripple-free dc in all other locations. See also section {8 .17.3}.

3 Protection by functional extra-low voltage (FELV)

This system uses the same safe voltage levels as SELV, but not all the protective measures required for SELV are needed and the system is widely used for supplies to power tools on construction sites {8.5}. The voltage must not exceed 50 V ac or 120 V dc. The reason for the difference is partly that direct voltage is not so likely to produce harmful shock effects in the human body as alternating current, and partly because the stated value of alternating voltage is r.m.s. and not maximum. As {Fig 3.9} shows, such a voltage rises to a peak of nearly 71 V, and in some circumstances twice this voltage level may be present. The allowable 120 V dc must be ripple free. See {8 .17.4}.

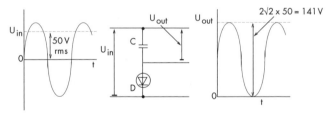

Fig 3.9 An alternating supply of 50 V may provide 141 V when the supply is rectified

{Figure 3.10} shows how a 120 V direct voltage with an 80 V peak-to-peak ripple will give a peak voltage of 160 V. The allowable ripple is such that a 120 V system must never rise above 140 V or a 60 V system above 70 V. It is interesting to note that a direct voltage with a superimposed ripple is more likely to cause heart fibrillation {3.4.2} than one which has a steady voltage. Unlike the SELV system, functional extra low voltage supplies are earthed as a normal installation. Direct contact is prevented by enclosures giving protection to IP2X (which means that live parts cannot be touched from outside by a human finger – (see {Table 2.4}) or by insulation capable of withstanding 500 V r.m.s. a.c. for one minute.

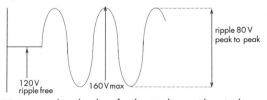

Fig 3.10 Increased peak value of a direct voltage with a ripple

It must be quite impossible for the low voltage levels of the normal installation to appear on the SELV system, and enclosures/insulation used for their separation must be subjected to the same insulation resistance tests as for the higher voltage. Any plugs used in such a circuit must not be interchangeable with those used on the higher voltage system. This will prevent accidentally applying a low voltage to an extra low voltage circuit. See also section {8 .17.4}.

4 Protection by limitation of discharge energy
Most electricity supply systems are capable of providing more than enough energy to cause death by electric shock. In some cases, there is too little energy to cause severe damage. For example, most electricians will be conversant with the battery-operated insulation resistance tester. Although the device operates at a lethal voltage (seldom less than 500 V dc) the battery is not usually capable of providing enough energy to give a fatal shock. In addition, the internal resistance of the instrument is high enough to cause a volt drop that reduces supply voltage to a safe value before the current reaches a dangerous level. This does not mean that the device is safe: it can still give shocks that may result in dangerous falls or other physical or mental problems. The electric cattle fence is a very good example of a system with limited energy. The system is capable of providing a painful shock to livestock, but not of killing the animals, which are much more susceptible to the effects of shock than humans.

3.4.5 Basic protection [411.8.2, 413, 414]
In the 16th Edition of the Regulations this was called 'direct contact protection'. The methods of preventing direct contact are mainly concerned with making sure that people cannot touch live conductors. These methods include:
1 the insulation of live parts – this is the standard method. The insulated conductors may be further protected by sheathing, conduit, etc.
2 the provision of barriers, obstacles or enclosures to prevent touching (IP2X). Where surfaces are horizontal and accessible, IP4X protection (solid objects wider than 1 mm are excluded – see {Table 2.4}), applies
3 placing out of reach or the provision of obstacles to prevent people from reaching live parts. Placing out of reach should not be used in locations of increased shock risk, and barriers must not be used except where the area is accessible only to skilled or instructed persons.
4 the provision of residual current devices (RCDs) provides supplementary protection {5.9} but only when contact is from a live part to an earthed part.

3.4.6 Fault protection [411.8.3, 413, 414]
In the 16th Edition of the Regulations this was called indirect contact protection. There are three methods of providing protection from shock after contact with a conductor that would not normally be live:
1 making sure that when a fault occurs and makes the parts live, it results in the supply being cut off within a safe time. In practice, this involves limitation of earth fault loop impedance, a subject dealt with in greater detail in {5.3}.
2 cutting off the supply before a fatal shock can be received using a residual current device {5.9}.
3 applying local supplementary equipotential bonding which will ensure that the resistance between parts which can be touched simultaneously is so low

that it is impossible for a dangerous potential difference to exist between them. It is important to stress that whilst this course of action will eliminate the danger of indirect contact, it will still be necessary to provide disconnection of the supply to guard against other faults, such as overheating.

It is important to appreciate that in some cases a dangerous voltage may be maintained if an uninterruptible power supply (UPS) or a standby generator with automatic starting is in use.

3.4.7 Protection for users of equipment outdoors [431]

Because of the reduced resistance to earth which is likely for people outdoors, special requirements apply to portable equipments used by them. Experience shows that many accidents have occurred whilst people are using lawn mowers, hedge trimmers, etc. There must be a suitable plug (a weather-proof type if installed outdoors) installed in a position that will not result in the need for flexible cords of excessive length. The plug must be protected by an RCD with a rating no greater than 30 mA that will trip in no more than 40 ms when an earth current of 150 mA flows. Where equipment is fed from a transformer (eg the centre-tapped 110 V system on construction sites), the transformer primary must be fed through a 30 mA rated RCD. These requirements do not apply to extra-low voltage sockets installed for special purposes, such as exterior low voltage garden lighting (see {8.15.2}).

3.5 HIGH TEMPERATURE PROTECTION

3.5.1 Introduction [131.3, 42]

The Regulations are intended to prevent both fires and burns which arise from electrical causes. Equipment must be selected and installed with the prevention of fire and burns fully considered. Three categories of thermal hazard are associated with an electrical installation.
  1 ignition arising directly from the installation,
  2 the spread of fire along cable runs or through trunking where proper fire stops have not been provided, and
  3 burns from electrical equipment.

The heat from direct sunlight will add significantly to the temperature of cables, and 20°C must be added to the ambient temperature when derating a cable subject to direct sunlight, unless it is permanently shaded in a way that does not reduce ventilation. Account must also be taken of the effect of the ultra-violet content of sunlight on the sheath and insulation of some types of cable.

Some types of electrical equipment are intended to become hot in normal service, and special attention is needed in these cases. For example, electric surface heating systems must comply fully with all three parts of BS 6351. Part 1 concerns the manufacture and design of the equipment itself, Part 2 with the design of the system in which it is used, and Part 3 its installation, maintenance and testing.

3.5.2 Fire protection [421, 559]

Where an electrical installation or a piece of equipment which is part of it is, under normal circumstances, likely to become hot enough to set fire to material close to it, it must be enclosed in heat and fire resistant material which will prevent danger. Because of the complexity of the subject, the Regulations give no specific guidance concerning materials or clearance dimensions. It is left to the designer to take account of the circumstances arising in a particular situation. When fixed equipment is chosen by the installation user or by some other party than the designer or the installer, the latter are still responsible for ensuring that the installation requirements of the manufacturers are met.

The same general principle applies in cases where an equipment may emit arcs or hot particles under fault conditions, including arc welding sets. Whilst it may be impossible in every case to prevent the outbreak of fire, attention must be paid to the means of preventing its spread.

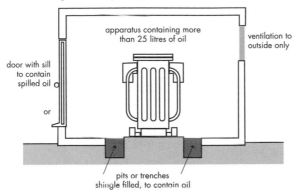

Fig 3.11 Precautions for equipment which contains flammable liquid

For example, any equipment which contains more than 25 litres of flammable liquid, must be so positioned and installed that burning liquid cannot escape the vicinity of the equipment and thus spread fire. {Figure 3.11} indicates the enclosure needed for such a piece of equipment, for example an oil-filled transformer. In situations where fire or smoke could cause particular hazards, consideration should be given to the use of low smoke and fire (LSF) cables. Such cables include those with thermosetting insulation and mineral-insulated types.

Perhaps a word is needed here concerning the use of the word 'flammable'. It means something that can catch fire and burn. We still see the word inflammable in everyday use, with the same meaning as flammable. This is very confusing, because the prefix 'in' may be taken as meaning 'not', giving exactly the opposite meaning. 'Inflammable' should never be used, 'nonflammable' being the correct term for something that cannot catch fire.

Under some conditions, especially where a heavy current is broken, the current may continue to flow through the air in the form of an arc. This is more likely if the air concerned is polluted with dust, smoke, etc. The arc will be extremely hot and is likely to cause burns to both equipments and to people; metal melted by the arc may be emitted from it in the form of extremely hot particles that will themselves cause fires and burns unless precautions are taken. Special materials that are capable of withstanding such arc damage are available and must be used to screen and protect surroundings from the arc.

Some types of electrical equipment, notably spotlights and halogen heaters, project considerable radiant heat. The installer must consider the materials that are subject to this heat to ensure that fire will not occur. Enclosures of electrical equipment must be suitably heat-resisting. Recessed or semi-recessed luminaires mounted in ceiling voids must be given special attention to ensure the heat they produce cannot result in fire by the installation of a non-flammable cowl above each of them. Equipment that focuses heat, such as radiant heaters and some luminaires, must be mounted so that excessive temperatures are not reached in adjacent surfaces. The installation of a protecting RCD with a rating not exceeding 300 mA will sometimes prevent a fire in the event of an earth fault.

Additions to an installation or changes in the use of the area it serves may give rise to fire risks. Examples are the addition of thermal insulation, the installation of additional cables in conduit or trunking, dust or dirt that restricts ventilation openings or forms an explosive mixture with air, changing lamps for others of higher rating, missing covers on joint boxes and other enclosures so that vermin may attack cables, and so on. (See also {3.9}).

3.5.3 Protection from burns [423]

The Regulations provide a Table showing the maximum allowable temperatures of surfaces which could be touched and thus cause burns. The allowable temperature depends on whether the surface is metallic or nonmetallic, and on the likely contact between the hand and the surface. Details follow in {Table 3.1}.

Table 3.1 Allowable surface temperatures for accessible parts
(taken from [Table 42.1] of BS 7671: 2008)

Part	Surface material	Max. temp (°C)
Hand held	metallic	55
	non-metallic	65
May be touched but not held	metallic	70
	non-metallic	80
Need not be touched in normal use	metallic	80
	non-metallic	90

The elements of forced air heaters must not be switched on until the rate of air flow across them is sufficient to ensure that the air emitted is not too hot, and water heaters and steam raisers must be provided with non self-resetting controls where appropriate. The suitability for connection of high temperature cables must be established with the manufacturer before cables running at more than 70°C are connected. Special attention must be paid to the likely temperature of hot surfaces where they may be touched by the very young, very old or the infirm.

3.6 OVERCURRENT PROTECTION

3.6.1 Introduction [43]

'Overcurrent' means what it says – a greater level of current than the materials in use will tolerate for a long period of time. The protection considered here will not necessarily apply to the equipment fed or to flexible cords or cables connecting such equipment to plugs and socket outlets. The term can be divided into two types of excess current.

1 Overload currents

These are currents higher than those intended to be present in the system. If such currents persist they will result in an increase in conductor temperature, and hence a rise in insulation temperature. High conductor temperatures are of little consequence except that the resistance of the conductor will be increased leading to greater levels of voltage drop.

Insulation cannot tolerate high temperatures since they will lead to deterioration and eventually failure. The most common insulation material is p.v.c. If it becomes too hot it softens, allowing conductors which press against it (and this will happen in all cases where a conductor is bent) to migrate through it so that they come close to, or even move beyond, the insulation surface. For this reason, p.v.c. insulation should not normally run at temperatures higher than 70°C, whereas under overload conditions it may have allowable temperatures up to 115°C for a short period during transient conditions.

2 Short circuit currents

These currents will only occur under fault conditions, and may be very high indeed. As we shall shortly show (see {3.6.3 and 3.6.4}) such currents will open the protective devices very quickly. These currents will not flow for long periods, so that under such short-term circumstances the temperature of p.v.c. insulation may be allowed to rise to 160°C. The clearance time of the protective device is governed by the adiabatic equation which is considered more fully in {3.7.3}.

3.6.2 Overload [432.2, 433]

Overload currents occur in circuits which have no faults but are carrying a higher current than the design value due to overloaded machines, an error in the assessment of diversity, and so on. When a conductor system carries more current than its design value, there is a danger of the conductors, and hence the insulation, reaching temperatures which will reduce the useful life of the system.

The devices used to detect such overloads, and to break the circuit for protection against them, fall into three main categories:

1 Semi-enclosed (rewirable) fuses to BS 3036 and cartridge fuses for use in plugs to BS 1362.
2 High breaking capacity (HBC) fuses to BS 88 and BS 1361. These fuses are still often known as high rupturing capacity (HRC) types.
3 Circuit breakers, miniature and moulded case types to BS EN 60898.

Examination of the characteristics of these devices {Figs 3.13 to 3.18} indicates that they are not the 'instant protectors' they are widely assumed to be. For example, an overloaded 30 A semi-enclosed fuse takes about 200 s to 'blow' when carrying twice its rated current. If it carries 450 A in the event of a fault (fifteen times rated current), it takes about 0.1 s to operate, or five complete cycles of a 50 Hz supply.

HBC fuses are faster in operation, but BS 88 Part 2 specifies that a fuse rated at 63 A or less must NOT operate within one hour when carrying a current 20% greater than its rating. For higher rated fuses, operation must not be within four hours at the same percentage overload. The latter are only required to operate within four hours when carrying 60% more current than their rated value.

Circuit breakers are slower in operation than is generally believed. For example, BS EN 60898 only requires a 32 A miniature circuit breaker to operate within one hour when carrying a current of 40 A. A 32 A type B MCB will take 5 s to open when carrying a current of 160 A. At very high currents operation is described by the BS as 'instantaneous' which is actually within 0.01 seconds.

All protective devices, then, will carry overload currents for significant times without opening. The designer must take this fact into account in his calculations. The circuit must be designed to prevent, as far as possible, the presence of comparatively small overloads of long duration.

The overload provisions of the Regulations are met if the setting of the device:

1 exceeds the circuit design current

2 does NOT exceed the rating of the smallest cable protected

In addition, the current for operation must not be greater than 1.45 times the rating of the smallest cable protected.

The overload protection can be placed anywhere along the run of a cable provided there are no branches, or must be at the point of cable size reduction where this occurs. There must be NO protection in the secondary circuit of a current transformer, or other situation where operation of the protective device would result in greater danger than that caused by the overload. Fuses and circuit breakers controlling a small installation are commonly grouped in a consumer's unit at the mains position.

There are some circuits that have widely varying loads, and it would be unfortunate if the protection operated due to a severe but short-lived overload. In such cases, the heating effect of the currents must be taken into account so that the overload setting is based on the thermal loading.

3.6.3 Fuses [Appendix 3]

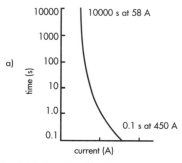
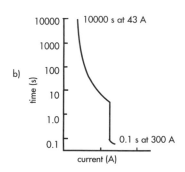

Fig 3.12 Time/current characteristics
a) 30 A semi-enclosed fuse b) 32 A miniature circuit breaker type B

Fuses operate because the fuse element is the 'weak link' in the circuit, so that overcurrent will melt it and break the circuit. The time taken for the fuse link to break the circuit (to 'blow') varies depending on the type of fuse and on the characteristic of the device. The time/current characteristic of a typical fuse is shown in {Fig 3.12(a)}. Curves for other types and ratings of fuses are shown in {Figs 3.13 to 3.15}. The figures are adapted from Appendix 3 of the BS 7671: 2008.

Where the current carried is very much greater than the rated value (which is usually associated with a fault rather than with an overload) operation is usually quite fast. For small overloads, where the current is not much larger than the rated value, operation may take a very long time, as indicated.

A graph with linear axes would need to be very large indeed if the high current/short time and the low current/long time ends of the characteristic were to be used to read the time to operate for a given current. The problem is removed by using logarithmic scales, which open out the low current and short time portions of the scales, and compress the high current and long time portions.

This means that the space between two major lines on the axes of the graph represents a change of ten times that represented by the two adjacent lines. In other words, a very much-increased range of values can be accommodated on a graph of a given size.

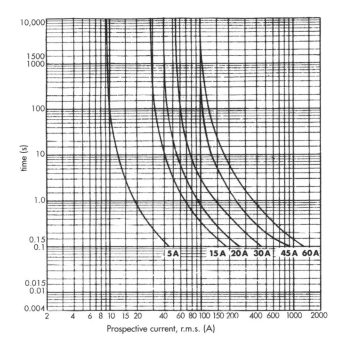

Fig 3.13 Time/current characteristics of semi-enclosed fuses to BS 3036

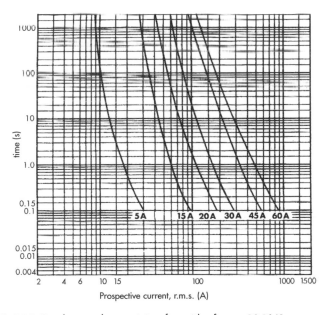

Fig 3.14 Time/current characteristics of cartridge fuses to BS 1361

Rewirable (semi-enclosed) fuses to BS 3036 may still be used, but as they can easily have the wrong fuse element (fuse wire) fitted and have low breaking capacity {3.7.2} they are not recommended. Where used, they are subject to the derating requirements which are explained in {4.3.8}.

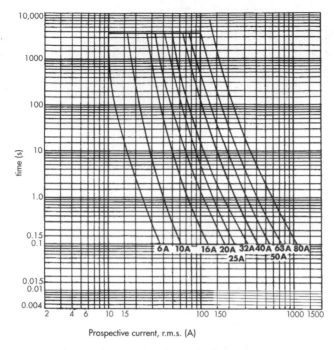

Prospective current, r.m.s. (A)

Fig 3.15 Time/current characteristics of cartridge fuses to BS 88 Part 2

All fuses must be clearly labelled with the fuse rating to make replacement with the wrong fuse as unlikely as possible. It must not be hazardous to make or break a circuit by insertion or removal of a fuse.

3.6.4 Circuit breakers [Appendix 3]

Circuit breakers operate using one or both of two principles. They are:

1 Thermal operation relies on the extra heat produced by the high current warming a bimetal strip, which bends to trip the operating contacts.

2 Magnetic operation is due to the magnetic field set up by a coil carrying the current, which attracts an iron part to trip the breaker when the current becomes large enough.

Thermal operation is slow, so it is not suitable for the speedy disconnection required to clear fault currents. However, it is ideal for operation in the event of small but prolonged overload currents. Magnetic operation can be very fast and so it is used for breaking fault currents; in many cases, both thermal and magnetic operation are combined to make the circuit breaker more suitable for both overload and fault protection. It must be remembered that the mechanical operation of opening the contacts takes a definite minimum time, typically 20 ms, so there can never be the possibility of truly instantaneous operation. The time/current graphs for miniature circuit breakers only show data for operating times down to 0.1 s (100 ms) which is stated to be 'instantaneous operation' (see {Figs 3.16 to 3.18}). Should an installation designer need information on faster operation (which seems unlikely) he must obtain it from the circuit breaker manufacturer. A typical time/current characteristic for a circuit breaker is shown {Fig. 3.12(b)}

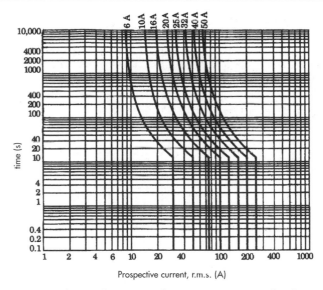

Prospective current, r.m.s. (A)

Fig 3.16 Time/current characteristics for some miniature circuit breakers Type B

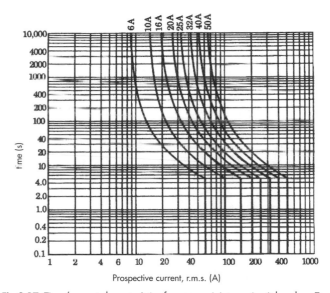

Prospective current, r.m.s. (A)

Fig 3.17 Time/current characteristics for some miniature circuit breakers Type C

All circuit breakers must have an indication of their current rating. Miniature circuit breakers have fixed ratings but moulded case types can be adjusted. Such adjustment must require the use of a key or a tool so that the rating is unlikely to be altered except by a skilled or instructed person. There are many types and ratings of moulded case circuit breakers, and if they are used, reference should be made to supplier's literature for their characteristics. Miniature circuit breakers are manufactured in fixed ratings from 6 A to 125 A and in three types, type B giving the closest protection. Operating characteristics for some of the more commonly used ratings of types B, C and D are shown in {Figs 3.16 to 3.18}. Short circuit current ratings for types B, C and D will be a minimum of 3 kA and may be as high as 25 kA.

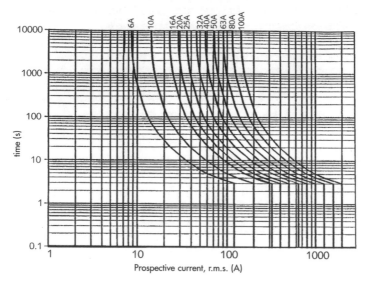

Fig 3.18 Time/current characteristics for some miniature circuit breakers Type D

The time/current characteristics of all circuit breakers {Figs 3.16 to 3.18} have a vertical section where there is a wide range of operating times for a certain current. Hence, with a fixed supply voltage, the maximum earth fault loop impedance is also fixed over this range of time. The operating current during the time concerned is a fixed multiple of the rated current. For example, a Type B MCB has a multiple of 5 (from {Table 3.2}) so a 32 A device of this type will operate over the time range of 0.1 s to 12 s at a current of 5 x 32 A = 160 A.

Table 3.2 Operating time ranges and current multiples for MCBs over fixed current section of characteristic
(from [Figs 3.4 to 3.6, Appendix 3] of BS 7671: 2008)

MCB Type	Range of operating times (s)	Current multiple of rating
B	0.04 to 13	x 5
C	0.04 to 5	x 10
D	0.05 to 3	x 20

Table 3.3 A comparison of types of protective device.

Semi-enclosed fuses	HBC fuses	Miniature circuit breakers
Very low initial cost	Medium initial cost	High initial cost
Low replacement cost	Medium replacement cost	Zero replacement cost
Low breaking capacity	Very high breaking capacity	Medium breaking capacity

Table 3.4 Comparison of miniature circuit breaker types

Type	Will not trip in 100 ms at rating	Will trip in 100ms at rating	Typical application
B	3 x	5 x	General purpose use (close protection)
C	5 x	10 x	Commercial and industrial applications with fluorescent fittings
D	10 x	50 x	Applications where high in-rush currents are likely (transformers, welding machines)

{Table 3.3} shows a comparison of the three main types of protective device in terms of cost, whilst {Table 3.4} compares the available types of MCB.

3.6.5 Protecting conductors [433]

The prime function of overload protection is to safeguard conductors and cables from becoming too hot. Thus the fuse or circuit breaker rating must be no greater than that of the smallest cable protected. Reference to the time/current characteristics of protective devices {Figs 3.13 to 3.18} shows that a significantly greater current than the rated value is needed to ensure operation.

Thus, the current at which the protective device operates must never be greater than 1.45 times the rating of the smallest cable protected. For example, consider a cable system rated at 30 A and protected by a miniature circuit breaker Type C, rated at 32 A. Reference to {Fig 3.17} shows that a prolonged overload of about 42 A will open the breaker after about 10 400 seconds (nearly three hours!). The ratio of operating current over rated current is thus 42/30 or 1.40, significantly lower than the maximum of 1.45. All circuit breakers and HBC fuses listed in {3.6.2 sections 2 and 3} will comply with the Regulations if their rating does not exceed that of the smallest cable protected.

Semi-enclosed (rewirable) fuses do not operate so closely to their ratings as do circuit breakers and HBC fuses. For example, the time/current characteristics of {Fig 3.13} show that about 53 A is needed to ensure the operation of a 30 A fuse after 10,000 s, giving a ratio of 53/30 or 1.77. For semi-enclosed fuses, the Regulations require that the fuse current rating must not exceed 0.725 times the rating of the smallest cable protected. Considering the 30 A cable protected by the 32 A miniature circuit breaker above, if a semi-enclosed fuse replaced the circuit breaker, its rating must not be greater than 0.725 x 30 or 21.8 A.

Since overload protection is related to the current-carrying capacity of the cables protected, it follows that any reduction in this capacity requires overload protection at the point of reduction. Reduced current-carrying capacity may be due to any one or more of:

1 a reduction in the cross-sectional area of the cable
2 a different type of cable
3 the cable differently installed so that its ability to lose heat is reduced
4 a change in the ambient temperature to which the cable is subjected
5 the cable is grouped with others.

{Figure 3.19} shows part of a system to indicate how protection could be applied to conductors with reduced current carrying capacity.

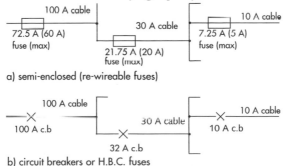

a) semi-enclosed (re-wireable fuses)

b) circuit breakers or H.B.C. fuses

Fig 3.19 Position and rating of devices for overload protection

In fact, the calculated fuse sizes for {Fig 3.19a} of 72.5 A, 21.75 A and 7.25 A are not available, so the next lowest sizes of 60 A, 20 A and 5 A respectively must be used. It would be unwise to replace circuit breakers with semi-enclosed fuses because difficulties are likely to arise. For example, the 5 A fuse used as the nearest practical size below 7.25 A is shown in {Fig 3.13} to operate in 100 s when carrying a current of 10 A. Thus, if the final circuit is actually carrying 10 A, replacing a 10 A circuit breaker with a 5 A fuse will result in the opening of the circuit. The temptation may be to use the next semi-enclosed fuse size of 15 A, but that fuse takes nearly seven minutes to operate at a current of 30 A. Clearly, the cable could well be damaged by excessive temperature if overloaded.

The device protecting against overload may be positioned on the load side of (downstream from) the point of reduction, provided that the unprotected cable length does not exceed 3 m, that fault current is unlikely, and that the cable is not in a position that is hazardous from the point of view of ignition of its surroundings. This Regulation is useful when designing switchboards, where a short length of cable protected by conduit or trunking feeds a low-current switch fuse from a high current fuse as in {Fig 3.22}.

All phase conductors must be protected, but attention must be paid to the need to break at the same time all three line conductors to a three-phase motor in the event of a fault on one phase, to prevent the motor from being damaged by 'single-phasing'. Normally the neutral of a three phase system should not be broken, because this could lead to high voltages if the load is unbalanced. Where the neutral is of reduced size, overload protection of the neutral conductor may be necessary, but then a circuit breaker must be used so that the phases are broken simultaneously.

Where unexpected operation of overload protection may cause danger or damage it may be omitted, but provision of an overload alarm should be considered (see {3.8.3}).

3.7 PROTECTION FROM FAULTS

3.7.1 Introduction [434]

The overload currents considered in the previous section are unlikely to be more than two to three times the normal rated current. Fault currents, on the other hand, can well be several hundreds, or even several thousands of times normal. In the event of a short circuit or an earth fault causing such current, the circuit must be broken before the cables are damaged by high temperatures or by electro-mechanical stresses. The latter stresses will be due to the force on a current carrying conductor which is subject to the magnetic field set up by adjacent conductors. This force is proportional to the current, and to the magnetic field strength. Since the field strength also depends on the current, force is proportional to the square of the current. If the current is one thousand times normal, force will be one million times greater than usual! Fault protection must not only be able to break such currents, but to do so before damage results. Abrasion of cable insulation by movement is usually prevented by normal fixings or by being enclosed in conduit or trunking. Support must be provided to cables in busbar chambers.

3.7.2 Prospective short-circuit current (PSC) [313.1, 434]

The current which is likely to flow in a circuit if line and neutral cables are short circuited is called the prospective short circuit current (PSC). It is the largest current that can flow in the system, and protective devices must be capable of breaking it safely. The breaking capacity of a fuse or of a circuit breaker is one of the factors that need to be considered in its selection. Consumer units to BS EN 60439-3 and BS 88 (HBC) fuses are capable of breaking any probable prospective short-circuit current, but before using other equipments the installer must make sure that their breaking capacity exceeds the PSC at the point at which they are to be installed.

The effective breaking capacity of over current devices varies widely with their construction. Semi-enclosed fuses are capable of breaking currents of 1 kA to 4 kA depending on their type, whilst cartridge fuses to BS 1361 will safely break at 16.5 kA for type I or 33 kA for type II. BS 88 fuses are capable of breaking any possible short-circuit current. Miniature circuit breakers to BS EN60898 have their rated breaking capacity marked on their cases in amperes (not kA) although above 10000 A the MCB may be damaged and lower breaking currents (75% for 10000 A and 50% above that level) must be used for design purposes.

Prospective short circuit current is driven by the e.m.f. of the secondary winding of the supply transformer through an impedance made up of the secondary winding and the cables from the transformer to the fault {Fig 3.20}. The impedance of the cables will depend on their size and length, so the PSC value will vary throughout the installation, becoming smaller as the distance from the intake position increases. [313.1] requires the PSC at the origin of the installation to be 'assessed by calculation, measurement, enquiry or inspection'. In practice, this can be difficult because it depends to some extent on impedances which are not only outside the installation in the supply system, but are also live. If the impedance of the supply system can be found, a straightforward calculation using the formula of {Fig 3.20} can be used, but this is seldom the case. An alternative is to ask the local Electricity Company. The problem here is that they are likely to protect themselves by giving a figure which is usually at least 16 kA in excess of the true value. The problem with using this figure is that the higher the breaking capacity of fuses and circuit breakers are (and this must never be less than the PSC for the point at which they are installed), the higher will be their cost. {Table 3.5} gives a method of assessing the PSC if the type and length of the service cable is known.

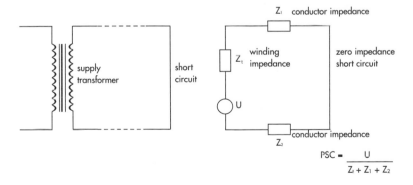

Fig 3.20 Prospective short circuit current (PSC)

Table 3.5 Estimation of PSC at the intake position

Length of supply (m)	PSC (kA) up to 25 mm² Al, 16 mm² Cu supply cable	PSC (kA) over 35 mm² Al (m) 25 mm² Cu supply cable
5	10.0	12.0
10	7.8	9.3
15	6.0	7.4
20	4.9	6.2
25	4.1	5.3
30	3.5	4.6
40	2.7	3.6
50	2.0	3.0

The table is not applicable in London or other major city centres, where the density of the distribution system means that higher values may apply. In this case it will be necessary to consult the supplier.

There are two methods for measuring the value of PSC, but these can only be used when the supply has already been connected. By then, the fuses and circuit breakers will already be installed.

The first method is to measure the impedance of the supply by determining its voltage regulation, that is, the amount by which the voltage falls with an increase in current. For example, consider an installation with a no-load terminal voltage of 230 V. If, when a current of 40 A flows, the voltage falls to 228 V, the volt drop will be due to the impedance of the supply.

$$\text{Thus } Z_s = \frac{\text{system volt drop}}{\text{current}} = \frac{230 - 228}{40} \, \Omega = \frac{2}{40} \, \Omega = 0.05 \, \Omega$$

$$\text{Then PSC} = \frac{U_o}{Z_s} = \frac{230}{0.05} \, \text{A} = 4600 \, \text{A or } 4.6 \, \text{kA}$$

A second measurement method is to use a loop impedance tester {see 5.3 and 7.6.2} connected to phase and neutral (instead of phase and earth) to measure supply impedance. This can then be used with the supply voltage as above to calculate PSC. Some manufacturers modify their earth-loop testers so that this connection is made by selecting 'PSC' with a switch. The instrument measures supply voltage, and calculates, then displays, PSC.

A possible difficulty in measuring PSC, and thus being able to use fuses or circuit breakers with a lower breaking capacity than that suggested by the Supply Company, is that the supply may later be reinforced. More load may result in extra or different transformers and cables being installed, which may reduce supply impedance and increase PSC.

3.7.3 Operating time [430.3]

Not only must the short-circuit protection system open the circuit to cut off a fault, but it must do so quickly enough to prevent both a damaging rise in the conductor insulation temperature and mechanical damage due to cable movement under the influence of electro-mechanical force. The time taken for the operation of fuses and circuit breakers of various types and ratings is shown in {Figs 3.13 to 3.18}. When the prospective short circuit current (PSC) for the point at which the protection is installed is less than its breaking capacity there will be no problem. When a short circuit occurs there will be a high current which must be interrupted quickly to prevent a rapid rise in conductor temperature.

The position is complicated because the rise in conductor temperature results in an increase in resistance which leads to an increased loss of energy and increased heating ($W = I^2Rt$), where W is the energy (J), I is the current (A), R is the resistance (Ω) and t is the time (s). The Regulations make use of the adiabatic equation which assumes that all the energy dissipated in the conductor remains within it in the form of heat, because the faulty circuit is opened so quickly. The equation is:

$$t = \frac{k^2S^2}{I^2}$$

where t = the time for fault current to raise conductor temperature to the highest permissible level (s)

k = a factor which varies with the type of cable
S = the cross-sectional area of the conductor (mm²)
I = the fault current value (A) – this will be the PSC

Some cable temperatures and values of k for common cables are given in {Table 3.6}.

Table 3.6 Cable temperatures and k values for copper cable
(from [Table 43.1] of BS 7671: 2008)

Insulation material	Initial temp. (°C)	Final temp. (°C)	k
70°C thermoplastic (gen. purpose p.v.c.)	70	160	115
90°C thermoplastic (p.v.c.)	90	160	100
60°C thermosetting (rubber)	60	200	141
90°C thermosetting (rubber)	90	250	143
Mineral (plastic covered or exposed to touch	70 (sheath)	160	115
Mineral (bare, not exp. to touch, not touching combustible material)	105 (sheath)	250	135
Mineral (bare, exposed to touch or touching combustible material)	105	250	115

As an example, consider a 10 mm² cable with p.v.c. insulation protected by a 40 A fuse to BS 88 Part 2 in an installation where the loop impedance between lines at the point where the fuse is installed is 0.12 Ω. If the supply is 400 V three phase, the prospective short circuit current (PSC) will be:

$$I = \frac{U_L}{Z} A = \frac{400}{0.12} A = 3.33 \text{ kA}$$

From {Table 3.6}, k = 115

$$t = \frac{k^2S^2}{I^2} = \frac{115^2 \times 10^2}{3333^2} s = 0.119 \text{ s}$$

{Figure 3.15} shows that a 40 A fuse to BS 88 Part 2 will operate in 0.1 s when carrying a current of 400 A. Since the calculated PSC at 3333 A is much greater than 400 A, the fuse will almost certainly clear the fault in less time than 0.1 s. As this time is less than that calculated by using the adiabatic equation (0.119 s) the cable will be unharmed in the event of a short circuit fault.

It is important to appreciate that the adiabatic equation applies to all cables, regardless of size. Provided that a protective device on the load side of a circuit has a breaking capacity equal to or larger than the PSC of the circuit then that circuit complies with the PSC requirements of the Regulations (see {Fig 3.21} and see also the note in {8.16.1} concerning the use of dual rated fuses for motor protection).

3.7.4 Conductors of reduced current-carrying capacity [433]

Short circuit protection must be positioned at every point where a reduction in cable-current carrying capacity occurs, as for overload protection {Fig 3.19}. However, if short circuit protection on the supply side of the point of reduction (for example, at the incoming mains position) has a characteristic that protects the reduced conductors, no further protection is necessary {Fig 3.22}.

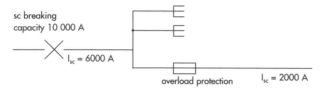

Fig 3.21 Short circuit protective device protecting a circuit of reduced cross-sectional area.

Even if not protected by a suitable device on the supply side, short circuit protection may be positioned on the load side of the reduction in rating if the conductors do not exceed 3 m in length, and are protected by trunking and conduit, and are not close to flammable materials. This reduction is particularly useful when connecting switchgear {Fig 3.22}. It should be noted that the 'tails' provided to connect to the supply system should always be of sufficient cross-sectional area to carry the expected maximum demand, and should never be smaller than 25 mm² for live conductors and 16 mm² for the main earthing conductor.

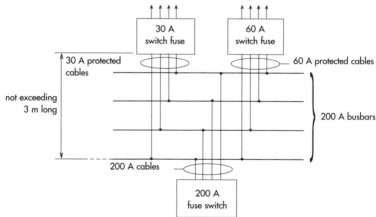

Fig 3.22 Short-circuit protection not required for short switchgear connections

3.7.5 Back-up protection [435, 536.4]

There are times when the overload protection has insufficient breaking capacity safely to interrupt the prospective short circuit current at the point of the installation where it is situated. An instance would be where a large number of low-rated miniature circuit breakers each with a breaking capacity of 3 kA are fed by a large cable.

In the event of a short circuit which gives a current of 8 kA, there is a good chance that the miniature circuit breaker concerned will be unable to break the fault.

Perhaps the fault current may continue to flow in the form of an arc across the opened circuit breaker contacts, causing a very high temperature and the danger of fire. Of course, if this happened the circuit breaker would be destroyed. The normal method of protection is to 'back up' the circuit with a protective device which has the necessary breaking capacity. For example, the group of miniature circuit breakers mentioned above could be backed up by an HBC fuse as shown in {Fig 3.23}.

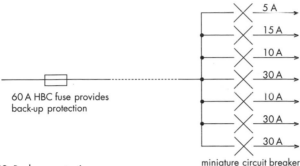

60 A HBC fuse provides
back-up protection

miniature circuit breaker

Fig 3.23 Back-up protection

It will be appreciated that when a protective device operates it does not do so instantaneously, and fault current will flow through it to the circuit it seeks to protect. The time for which such a current flows is a critical factor in the damage that may be done to the system before the fault clears. Damage applies to cables, switchgear, protective devices, etc. if the fault is not cleared quickly enough. This damage will be the result of the release of the energy of the fault current, and the system designer will aim to minimise it by calculation and by consulting manufacturers' data of energy let-through.

3.7.6 Insulation monitoring
In some special applications, usually IT systems, the insulation effectiveness of an installation or of an item of equipment is monitored continuously. This is done by the injection of a signal, usually at less than 50V, which measures insulation resistance. When this resistance falls below a preset level an alarm is triggered and/or the circuit concerned is switched off automatically. The setting of the monitor must be arranged so that it can only be altered by an authorised person, and to that end it must be possible to alter the setting only with a special tool.

3.8 SHORT CIRCUIT AND OVERLOAD PROTECTION
3.8.1 Combined protection [435.1]
In many practical applications, most types of fuses and circuit breakers are suitable for both overload and short circuit protection. Care must be taken, however, to ensure that the forms of protection are chosen so that they are properly coordinated to prevent problems related to excessive let-through of energy or to lack of discrimination {3.8.6}.

3.8.2 Current limited by supply characteristic [436]
If a supply has a high impedance, the maximum current it can provide could be less than the current carrying capacity of the cables in the installation. In such a case, no

overload or short circuit protection is required. Typical examples are bell transformers, and some welding sets. This situation is very unlikely with a supply taken from an Electricity Company, but could well apply to a private generating plant.

3.8.3 Protection omitted [434.3]

There are cases where a break in circuit current due to operation of a protective device may cause more danger than the overload or fault. For example, breaking the supply to a lifting electromagnet in a scrap yard will cause it to drop its load suddenly, possibly with dire consequences. If the field circuit of a dc motor is broken, the reduction in field flux may lead to a dangerous increase in speed. A current transformer has many more secondary than primary turns, so dangerously high voltages will occur if the secondary circuit is broken.

In situations like these the installation of an overload alarm will give warning of the faulty circuit, which can be switched off for inspection when it is safe to do so. The possibility of short circuits in such cables will be reduced if they are given extra protection. Probably the most usual case of omission of protection is at the incoming mains position of a small installation. Here, the supply fuse protects the installation tails and the consumer's unit. The unprotected equipment must, however, comply with the requirements for otherwise unprotected systems listed in {3.7.4}.

3.8.4 Protection of conductors in parallel [434.4, Appendix 10, Appendix 15]

The most common application of cables in parallel is in ring final circuits for socket outlets, whose special requirements will be considered in {6.3.2}. Cables may otherwise be connected in parallel provided that they are of exactly the same type, run together throughout their length, have no branches and are expected to share the total circuit current in proportion to their cross-sectional area. It is not recommended that cables are connected in parallel at all except in ring final circuits.

Overload protection must then be provided for the sum of the current-carrying capacities of the cables. If, for example, two cables with individual current ratings of 13 A are connected in parallel, overload protection must be provided for 26 A.

Account must also be taken of a short circuit that does not affect all the cables: this is made less likely by the requirement that they should run close together. In {Fig 3.24}, for example, the 30 A cable with the short circuit to neutral must be able to carry more than half the short circuit current without damage until the protection opens.

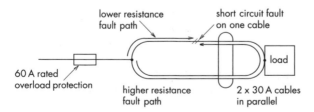

Fig 3.24 Cables in parallel. The lower resistance path will carry the higher fault current

Fault current sharing in these circumstances depends on the inverse of the ratio of the conductor resistances, or impedances where the cables are armoured. If, for example, the fault on one cable were to occur close to the connection to the protective device, almost all of the fault current would be carried by the short length from the protection to the fault. In these circumstances there would be little protection for the faulty cable, and it would be prudent to provide protection with the installation of a suitable RCD. Where heavy armoured cables are connected in parallel, unequal load sharing can result from the positions in which such cables are situated. See [Appendix 10] for detailed treatment. The same Appendix provides detailed information on the protection needed if three or more cables are connected in parallel.

3.8.5 Absence of protection [436]
The protection described in this Chapter applies to conductors, but not necessarily to the equipment fed. This is particularly true where flexible cords are concerned. For example, a short circuit in an appliance fed through a 0.5 mm² flexible cord from a 13 A plug may well result in serious damage to the cord or the equipment before the fuse in the plug will operate.

3.8.6 Selectivity or Discrimination [536]
Most installations include a number of protective devices in series, and they must operate correctly relative to each other if healthy circuits are not to be disconnected. 'Selectivity' has been used to replace the original word 'discrimination' for this effect. Selectivity occurs when the protective device nearest to the fault operates, leaving all other circuits working normally.

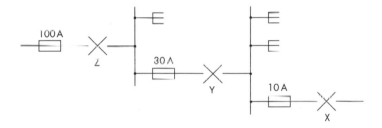

Fig 3.25 System layout to explain selectivity (discrimination)

{Figure 3.25} shows an installation with a 100 A main fuse and a 30 A submain fuse feeding a distribution board containing 10 A fuses. If a fault occurs at point Z, the 100 A fuse will operate and the whole installation will be disconnected. If the fault is at X, the 10 A fuse should operate and not the 30 A or 100 A fuses. A fault at Y should operate the 30 A, and not the 100 A fuse. If this happens, the system has selected (discriminated) properly.

Lack of selectivity would occur if a fault at X caused operation of the 30 A or 100 A fuses, but not the 10 A fuse. This sounds impossible until we remember the time/current fuse characteristics explained in {3.6.3}. For example, {Fig 3.26} shows the superimposed characteristics of a 5 A semi-enclosed fuse and a 10 A miniature circuit breaker which we shall assume are connected in series. If a fault current of 50 A flows, the fuse will operate in 0.56 s whilst the circuit breaker would take 24 s to open. Clearly the fuse will operate first and the devices have discriminated. However, if the fault current is 180 A, the circuit breaker will open in 0.016 s, well before the fuse would operate, which would take 0.12 s. In this case, there has been no discrimination.

To ensure selectivity is a very complicated matter, particularly where an installation includes a mixture of types of fuse, or of fuses and circuit breakers. Manufacturers' operating characteristics must be studied to ensure discrimination. As a rule of thumb where fuses or circuit breakers all of the same type are used, there should be a doubling of the rating as each step towards the supply is taken.

When fault current is high enough to result in operation of the protective device within 40 ms (two cycles of a 50 Hz supply), the simple consideration of characteristics as shown in Fig. 3.26 may not always result in correct discrimination and device manufacturers should be consulted.

When RCDs are connected in series, discrimination between them is also important, the rule here being that a trebling in rating applies with each step towards the supply (see also {5.9.3}).

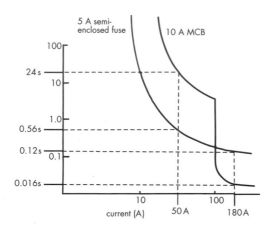

Fig 3.26 To illustrate a lack of selectivity (discrimination)

3.9 PROTECTIVE MEASURES AGAINST FIRE
3.9.1 Introduction [42]
Chapter 42 of the Regulations gives requirements to prevent electrical installations from causing danger or damage to people or livestock due to thermal radiation, smoke or fire and from causing safety services to be cut off due to failure of electrical equipment.

3.9.2 What locations are involved? [421, 422]
These Chapters regarding precautions against fire apply to three types of installation.

1 Installations in locations mainly constructed of combustible materials. An example could be a building largely made of wood.
2 Installations where there is a risk of fire due to the nature of the materials present. Examples are where combustible materials are being stored or processed and where there is likely to be a build-up of dust or fibre, such as in paper mills, textile factories, etc.
3 Installations in situations where evacuation in the event of fire would be difficult. Escape routes in such situations must be wired with systems that will resist fire for two hours.

3.9.3 Locations with combustible construction materials [422.4, Appendix 5]

The basis of these Regulations is to make sure that the electrical installation will not cause ignition of the structure. To this end, all electrical equipment, such as installation boxes, distribution boards, etc., must be suitable for the situation in which they are installed. Where they are not so, they must be enclosed in 12 mm of non-flammable material, such as fibreglass, or embedded in 100 mm of glass or mineral wool. Using such materials will, of course, reduce the effectiveness of heat dissipation, and this must be taken into account (see {4.3.6}). These requirements also apply to non-combustible walls which have combustible materials incorporated for thermal and sound insulation.

Claw fixings must not be used to secure equipment in this situation. All cables, cords, conduit and trunking must fully meet the minimum requirements of the relevant standards. Equipment such as luminaries must be kept at a safe distance from combustible materials as in item 1. of {3.9.4}.

3.9.4 Locations with fire risk due to the materials present [422.3]

This Section includes Regulations aimed at the prevention of ignition of combustible material stored in or processed in the location. Only electrical systems absolutely necessary for use in those locations should be installed there. Some locations have a higher risk of fire and of explosion due to the nature of processed and stored materials. In such situations, the following additional requirements apply

1 Luminaires must be kept at a safe distance from combustible materials. The distances depend on the rating of the light source and minimum values are 0.5 m for 100 W, 0.8 m for 300 W and 1.0 m for 500 W sources

2 The enclosures of electrical equipment must not exceed 90°C under normal conditions or 115°C under fault conditions.

3 Switchgear and control gear must be installed outside the area unless it is suitable for the location or has an enclosure rated at IP4X, or IP5X if dust is present

4 Unless the wiring system is completely embedded in non-combustible material such as plaster or concrete, it must have the flame propagation characteristics specified in BS EN 60332

5 A wiring system passing through the location but not feeding equipment within it must not employ bare conductors

6 Every luminaire must be suitable for the location, have an enclosure to IP5X, have a limited surface temperature and be of a type that prevents lamp components falling from it

7 Except for mineral insulated cables and busbar trunking systems, TN and TT systems must be protected by a 300 mA RCD, or a 30 mA RCD where a resistive fault could cause a fire. IT systems must have an insulation monitoring device with audible and visual signals and is subject to the requirements of item 4

8 A motor which is automatically or remotely controlled must have excessive temperature protection with manual reset

9 Overload and fault protective devices must be situated outside the location

10 Extra-low voltage circuits must have insulation capable of withstanding a test voltage of 500 V d.c. for 1 minute

11 PEN conductors must not be used unless they pass through the location and do not feed equipment within it

12 All circuits must be provided with means to isolate all live conductors, using a linked switch or a linked circuit breaker

13 Flexible cables and cords must be of the heavy duty type with a voltage rating of not less than 450/750 V, or must be suitably protected from mechanical damage

14 All heating appliances must be fixed, not portable

15 Heat storage appliances must be of a type that prevents the ignition of combustible dusts or fibres by the heat-storing core.

The same requirements apply to buildings where the construction materials are combustible or where the structure is likely to propagate fire. They also apply to locations of national, commercial, industrial or pubic significance, such as national monuments, museums and other public buildings. Other situations such as laboratories, computer centres and industrial and storage facilities may also qualify as being of commercial or industrial significance.

CHAPTER FOUR

Cables, conduits and trunking

4.1 INTRODUCTION [52]

This Chapter is concerned with the selection of wiring cables for use in an electrical installation. It also deals with the methods of supporting such cables, ways in which they can be enclosed to provide additional protection, and how the conductors are identified. All such cables must conform in all respects with the appropriate British Standard.

This Electrician's Guide does not deal with cables for use in supply systems, heating cables, or cables for use in the high voltage circuits of signs and special discharge lamps.

4.1.1 Cable insulation materials

Due to the ever-increasing number of materials available for conductor insulation, the Wiring Regulations has changed the names of the commoner materials. p.v.c. is no longer given that name, but is described as 'thermoplastic', defined as a material that will melt at a given temperature so that it can be moulded and reformed repeatedly. Rubber will now be called 'thermosetting'. It is a material that is chemically cross-linked (set in shape and not capable of being remoulded) and thus likely to perform better in certain high temperature conditions than a thermoplastic material. This removes the difficulty of classifying XLPE (cross-linked polyethylene) which is neither a plastic nor a rubber but is a thermosetting material. To help users, the British Standard still has the older terms in brackets, i.e. thermoplastic (p.v.c.) and thermosetting (rubber).

Rubber

For many years wiring cables were insulated with vulcanised natural rubber (VIR). Much cable of this type is still in service, although it is many years since it was last manufactured. Since the insulation is organic, it is subject to the normal ageing process, becoming hard and brittle. In this condition it will continue to give satis factory service unless it is disturbed, when the rubber cracks and loses its insulating properties. It is advisable that wiring of this type which is still in service should be replaced by a more modern cable. Synthetic thermosetting rubber compounds are used widely for insulation and sheathing of cables for flexible and for heavy duty applications. Many variations are possible, with conductor temperature ratings from 60°C to 180°C, as well as the ability to resist oil, ozone and ultra-violet radiation depending on the formulation.

Paper

Dry paper is an excellent insulator but loses its insulating properties if it becomes wet. Dry paper is hygroscopic, that is, it absorbs moisture from the air. It must be sealed to ensure that there is no contact with the air. Because of this, paper insulated cables are sheathed with impervious materials, lead being the most common. PILC (paper insulated lead covered) was traditionally used for heavy power work. The paper insulation is impregnated with oil or non-draining compound to improve its long-term performance. Cables of this kind needed special jointing methods to ensure that the insulation remains sealed. This difficulty, as well as the weight of the cable, has led to the widespread use of p.v.c. (thermoplastic) and XLPE (thermosetting) insulated cables in place of paper insulated types.

P.V.C.

Polyvinyl chloride (p.v.c.) is a thermoplastic material and is now the most usual low voltage cable insulation. It is clean to handle and is reasonably resistant to oils and other chemicals. When p.v.c. burns, it emits dense smoke and corrosive hydrogen chloride gas. The physical characteristics of the material change with temperature: when cold it becomes hard and difficult to strip, and so BS 7671 specifies that it should not be worked at temperatures below 5°C. However a special p.v.c. is available which remains flexible at temperatures down to -20°C.

At high temperatures the material becomes soft so that conductors that are pressing on the insulation (e.g. at bends) will 'migrate' through it, sometimes moving to the edge of the insulation. Because of this property the temperature of general purpose p.v.c. must not be allowed to exceed 70°C, although versions which will operate safely at temperatures up to 85°C are also available. If p.v.c. is exposed to sunlight it may be degraded by ultra-violet radiation. If it is in contact with absorbent materials, the plasticiser may be 'leached out' making the p.v.c. hard and brittle.

LSF (Low smoke and fume)

Materials that have reduced smoke and corrosive gas emissions in fire compared with p.v.c. have been available for some years. They are normally used as sheathing compounds over XLPE or LSF insulation, and can give considerable safety advantages in situations where numbers of people may have to be evacuated in the event of fire.

Thermosetting (XLPE)

Cross-linked polyethylene (XLPE) is a thermosetting compound that has better electrical properties than p.v.c. and is therefore used for medium- and high-voltage applications. It has more resistance to deformation at higher temperatures than p.v.c., which it is gradually replacing. It has also replaced PILC in some applications. Thermosetting insulation may be used safely with conductor temperatures up to 90°C, thus increasing the useful current rating, especially when ambient temperature is high. A LSF (low smoke and fume) type of thermosetting cable is available.

Mineral

Provided that it is kept dry, a mineral insulation such as magnesium oxide is an excellent insulator. Since it is hygroscopic (it absorbs moisture from the air) this insulation is kept sealed within a copper sheath. The resulting cable is totally fireproof and will operate at temperatures up to 250°C. It is also entirely inorganic and thus non-ageing These cables have small diameters compared with alternatives, great mechanical strength, are waterproof, resistant to radiation and electro-magnetic pulses, are pliable and corrosion resistant. In cases where the copper sheath may corrode, the cable is used with an overall LSF covering, which reduces the temperature at which the cable may be allowed to operate. Since it is necessary to prevent the ingress of moisture, special seals are used to terminate cables. Special mineral-insulated cables with twisted cores to reduce the effect of electro-magnetic interference are available.

4.2 CABLES

4.2.1 Non-flexible low voltage cables [521]

Types of cable currently satisfying the Regulations are shown in {Fig 4.1}.

Fig 4.1

a) Non-armoured p.v.c.-insulated cables

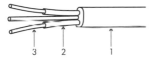

1 p.v.c sheath
2 p.v.c. insulation
3 copper conductor: solid, stranded or flexible

b) Armoured p.v.c.-insulated cables

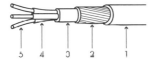

1 p.v.c. sheath
2 armour – galvanised steel wire
3 p.v.c. bedding
4 p.v.c. insulation
5 copper conductor

c) Split-concentric p.v.c.-insulated cables

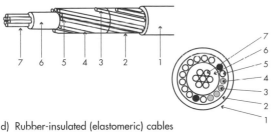

1 p.v.c. oversheath
2 Melinex binder
3 p.v.c. strings
4 neutral conductor: p.v.c.-covered wires
5 earth continuity conductor: bare copper wires
6 p.v.c. phase insulation
7 copper conductors

d) Rubber-insulated (elastomeric) cables

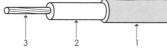

1 textile braided and compounded
2 85°C rubber insulation
3 tinned copper conductor

e) Impregnated-paper insulated lead sheathed cables

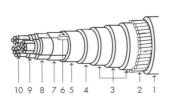

1 p.v.c. oversheath
2 galvanised steel wire armour
3 bedding
4 sheath: lead or lead alloy
5 copper woven fabric tape
6 filler
7 screen of metal tape intercalated with paper tape
8 impregnated paper insulation
9 carbon paper screen
10 shaped stranded conductor

f) Armoured cables with thermosetting insulation

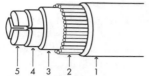

1 p.v.c. oversheath
2 galvanised steel wire armour
3 taped bedding
4 XLPE insulation
5 solid aluminium conductor

5 4 3 2 1

g) Mineral-insulated cables

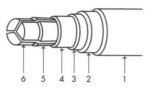

1 LSF oversheath
2 copper sheath
3 magnesium oxide insulation
4 copper conductors

4   3   2   1

h) Consac cables

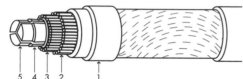

1 extruded p.v.c. or polythene oversheath
2 thin layer of bitumen containing a corrosion inhibitor
3 extruded smooth aluminium sheath
4 paper belt insulation
5 paper core insulation
6 solid aluminium conductors

6   5   4 3   2   1

i) Waveconal cables

1 extruded p.v.c. oversheath
2 aluminium wires
3 rubber anti-corrosion bedding
4 XLPE core insulation
5 solid aluminium conductors

5 4 3 2   1

[Table 52.1] gives the maximum conductor operating temperatures for the various types of cables. For general purpose p.v.c. this is 70°C. Cables with thermosetting insulation can be operated with conductor temperatures up to 90°C but since the accessories to which they are connected may be unable to tolerate such high temperatures, operation at 70°C is much more usual. Other values of interest to the electrician are shown in {Table 3.7}. Minimum cross-sectional areas for cables are shown in {Table 4.1}.

Table 4.1 Minimum permitted cross-sectional areas for cables
(from [Table 52.3] of BS 7671: 2008)

type of circuit	conductor material	cross-sectional area (mm²)
Power and lighting circuits (insulated conductors)	Copper	1.0
	Aluminium	16.0
Signalling and control circuits	Copper	0.5
Flexible, more than 7 cores	Copper	0.1*
Bare conductors and busbars	Copper	10.0
	Aluminium	16.0
Bare conductors for signalling and control	Copper	4.0

* 0.1 mm² is permitted for signalling and control circuits for electronic equipment

4.2.2 Cables for overhead lines

Any of the cables listed in the previous subsection are permitted to be used as overhead conductors provided that they are properly supported. Normally, of course, the cables used will comply with a British Standard referring particularly to special cables for use as overhead lines. Such cables include those with an internal or external catenary wire, which is usually of steel and is intended to support the weight of the cable over the span concerned.

Since overhead cables are to be installed outdoors, they must be chosen and erected so as to offset the problems of corrosion and ultra-violet radiation. Since such cables will usually be in tension, their supports must not damage the cable or its insulation. More information on corrosion is given in {4.2.5}.

4.2.3 Flexible low voltage cables and cords [521.9]

By definition flexible cables have conductors of cross-sectional area 4 mm² or greater, whilst flexible cords are sized at 4 mm² or smaller. Quite clearly, the electrician is nearly always concerned with flexible cords rather than flexible cables. {Figure 4.2} shows some of the many types of flexible cords that are available.

Fig 4.2 Flexible cords

a) Braided circular

1 oversheath – p.v.c.
2 braid – plain copper wire
3 inner sheath – p.v.c.
4 insulation – p.v.c. coloured
5 conductors – plain copper

b) Unkinkable

1 rubber layer collectively textile braided semi-embedded
2 insulation (cores) 60°C rubber
3 conductors – tinned copper

c) Circular sheathed

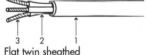

1 sheath – rubber or p.v.c.
2 insulation 60°C rubber or p.v.c.
3 conductors – tinned copper

d) Flat twin sheathed

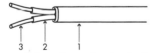

1 sheath – p.v.c.
2 insulation – p.v.c.
3 conductors – plain copper

e) Braided circular insulated with glass fibre

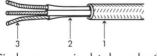

1 glass braided overall
2 insulation – silicon rubber
3 conductors – stranded copper

f) Single core p.v.c.-insulated non-sheathed

1 insulation – p.v.c.
2 conductors – plain copper

Flexible cords or cables should not normally be used for fixed wiring, but if they are, they must be visible throughout their length. The maximum mass that can be supported by each flexible cord is listed in [Table 4F3A], part of which is shown here as {Table 4.2}.

Table 4.2 Maximum mass supported by flexible cord
(from [Table 4F3A] of BS 7671: 2008)

Cross-sectional area (mm²)	Maximum mass to be supported (kg)
0.5	2
0.75	3
1.0	5
1.25	5
1.5	5

The temperature at the cord entry to luminaires is often very high, especially where filament lamps are used. It is important that the cable or flexible cord used for final entry is of a suitable heat resisting type, such as 150°C rubber-insulated and braided. {Fig 4.3} shows a short length of such a cord used to make the final connection to a luminaire.

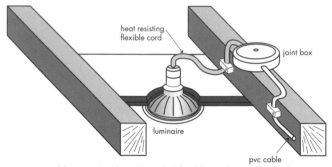

Fig 4.3 150°C rubber-insulated and braided flexible cord used for the final connection to a luminaire

4.2.4 Cables carrying alternating current [521.5]

Alternating current flowing in a conductor sets up an alternating magnetic field which is much stronger if the conductor is surrounded by an iron-rich material, for example if it is steel wire armoured or if it is installed in a steel conduit. The currents in a twin cable, or in two single core cables feeding a single load, will be the same. They will exert opposite magnetic effects that will almost cancel, so that virtually no magnetic flux is produced if they are both enclosed in the same conduit or armouring. The same is true of three-phase balanced or unbalanced circuits provided that all three (or four, where there is a neutral) cores are within the same steel armouring or steel conduit. An alternating flux in an iron core results in iron losses, which result in power loss appearing as heat in the metal enclosure. It should be remembered that not only will the heat produced by losses raise the temperature of the conductor, but that the energy involved will be paid for by the installation user through his electricity meter. Thus, it is important that all conductors of a circuit are contained within the same cable, or are in the same conduit if they are single-core types (see {Fig 4.4}).

A similar problem will occur when single-core conductors enter an enclosure through separate holes in a steel end plate {Fig 4.5}.

For this reason, single-core armoured cables should not be used. If the single core cable has a metal sheath which is non-magnetic, less magnetic flux will be produced. However, there will still be induced e.m.f. in the sheath, which can give rise to a circulating current and sheath heating.

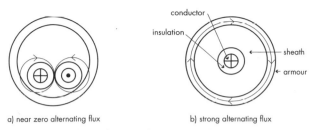

Fig 4.4 Iron losses in the steel surrounding a cable when it carries alternating current
a) twin conductors of the same single-phase circuit – no losses
b) single core conductor – high losses

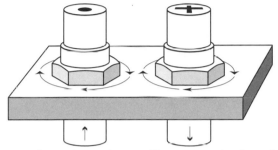

Fig 4.5 Iron losses when single-core cables enter a steel enclosure through separate holes

If mineral insulated cables are used, or if multi-core cables are used, with all conductors of a particular circuit being in the same cable, no problems will result. The copper sheath is non-magnetic, so the level of magnetic flux will be less than for a steel armoured cable; there will still be enough flux, particularly around a high current cable, to produce a significant induced e m f. However, multi-core mineral insulated cables are only made in sizes up to 25 mm² and if larger cables are needed they must be single core.

{Figure 4.6(a)} shows the path of circulating currents in the sheaths of such single core cables if both ends are bonded. {Figure 4.6(b)} shows a way of breaking the circuit for circulating currents.

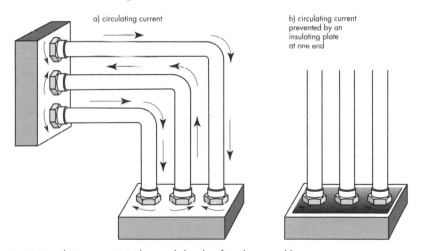

Fig 4.6 Circulating currents in the metal sheaths of single-core cables
a) bonded at both ends   b) circulating currents prevented by single point bonding

[523.10] calls for all single core cable sheaths to be bonded at both ends unless they have conductors of 70 mm² or greater. In that case they can be single point bonded if they have an insulating outer sheath, provided that:
  i) e.m.f. values no greater than 25 V to earth are involved, and
  ii) the circulating current causes no corrosion, and
  iii) there is no danger under fault conditions.

The last requirement is necessary because fault currents will be many times greater than normal load currents. This will result in correspondingly larger values of alternating magnetic flux and of induced e.m.f.

4.2.5 Corrosion [522.5]

The metal sheaths and armour of cables, metal conduit and conduit fittings, metal trunking and ducting, as well as the fixings of all these items, are likely to suffer corrosion in damp situations due to chemical or electrolytic attack by certain materials, unless special precautions are taken. The offending materials include:
  1 unpainted lime, cement and plaster,
  2 floors and dados including magnesium chloride,
  3 acidic woods, such as oak,
  4 plaster undercoats containing corrosive salts,
  5 dissimilar metals which will set up electrolytic action.

In all cases the solution to the problem of corrosion is to separate the materials between which the corrosion occurs. For chemical attack, this means having suitable coatings on the item to be installed, such as galvanising or an enamel or plastic coating. Bare copper sheathed cable, such as mineral insulated types, should not be laid in contact with galvanised material like a cable tray if conditions are likely to be damp. A p.v.c. covering on the cable will prevent a possible corrosion problem.

  To prevent electrolytic corrosion, which is particularly common with aluminium-sheathed cables or conduit, a careful choice of the fixings with which the aluminium comes into contact is important, especially in damp situations. Suitable materials are aluminium, alloys of aluminium which are corrosion resistant, zinc alloys complying with BS 1004, porcelain, plastics, or galvanised or sheradised iron or steel.

4.3 CABLE CHOICE

4.3.1 Cable types [521, 522, 527, 560.8]

When choosing a cable one of the most important factors is the temperature attained by its insulation (see {4.1.1}); if the temperature is allowed to exceed the upper design value, premature failure is likely. In addition, corrosion of the sheaths or enclosures may result. For example, bare conductors such as busbars may be operated at much higher temperatures than most insulated conductors.

  However, when an insulated conductor is connected to such a high temperature system, its own insulation may be affected by heat transmitted from the busbar, usually by conduction and by radiation. To ensure that the insulation is not damaged:
  either the operating temperature of the busbar must not exceed the safe
          temperature for the insulation,
  or   the conductor insulation must be removed for a suitable distance from
          the connection with the busbar and replaced with heat resistant
          insulation (see {Fig 4.7}).

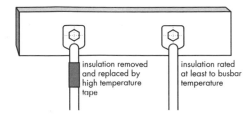

Fig 4.7 Insulation of a cable connected to a hot busbar

It is common sense that the cable chosen should be suitable for its purpose and for the surroundings in which it will operate. It should not be handled and installed in unsuitable temperatures. P.V.C. becomes hard and brittle at low temperatures, and if a cable insulated with it is installed at temperatures below 5°C it may well become damaged.

[522] includes a series of Regulations which are intended to ensure that suitable cables are chosen to prevent damage from temperature levels, moisture, dust and dirt, pollution, vibration, mechanical stress, plant growths, animals, sunlight or the kind of building in which they are installed. As already mentioned in {3.5.2}, cables must not produce, spread, or sustain fire.

[527] contains Regulations which are intended to reduce the risk of the spread of fire and are concerned with choosing cables with a low likelihood of flame propagation. A run of bunched cables is a special fire risk and cables in such a situation should comply with the standards stated above. Some cables must be able to continue to operate in a fire. These special cables are intended to be used when it is required to maintain circuit integrity for longer than is possible with normal cable. This is a very specialist field, a number of BS EN publications giving the requirements. See [560.8] for details.

4.3.2 Current carrying capacity of conductors [523, Appendix 4]
All cables have electrical resistance, so there must be an energy loss when they carry current. This loss appears as heat and the temperature of the cable rises. As it does so, the heat it loses to its surroundings by conduction, convection and radiation also increases. The rate of heat loss is a function of the difference in temperature between the conductor and the surroundings, so as the conductor temperature rises, so does its rate of heat loss.

A cable carrying a steady current, which produces a fixed heating effect, will get hotter until it reaches the balance temperature where heat input is equal to heat loss {Fig 4.8}. The final temperature achieved by the cable will thus depend on the current carried, how easily heat is dissipated from the cable and the temperature of the cable surroundings.

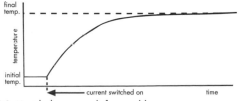

Fig 4.8 Heat balance graph for a cable

P.V.C. is probably the most usual form of insulation, and is very susceptible to damage by high temperatures. It is very important that p.v.c. insulation should not be allowed normally to exceed 70°C, so the current ratings of cables are designed to ensure that this will not happen. Some special types of p.v.c. may be used up to 85°C. A conductor temperature as high as 160°C is permissible under very short time fault conditions, on the assumption that when the the fault is cleared the p.v.c. insulation will dissipate the heat without itself reaching a dangerous temperature.

A different set of cable ratings will become necessary if the ability of a cable to shed its heat changes. Thus, [Appendix 4] has different Tables and columns for different types of cables, with differing conditions of installation, degrees of grouping and so on. For example, mineral insulation does not de-rate, even at very high temperatures. The insulation is also an excellent heat conductor, so the rating of such a cable depends on how hot its sheath can become rather than the temperature of its insulation.

For example, if a mineral insulated cable has an overall sheath of LSF or p.v.c., the copper sheath temperature must not exceed 70°C, whilst if the copper sheath is bare and cannot be touched and is not in contact with materials which are combustible its temperature can be allowed to reach 105°C. Thus, a 1mm^2 light duty twin mineral insulated cable has a current rating of 18.5 A when it has an LSF or p.v.c. sheath, or 22 A if bare and not exposed to touch. It should be noted that the cable volt drop will be higher if more current is carried (see {4.3.11}). [Appendix 4] includes a large number of Tables relating to the current rating of cables installed in various ways. The use of the Tables will be considered in more detail in {4.3.4 to 4.3.11}.

4.3.3 Methods of cable installation [521, 522, 528]

We have seen that the rating of a cable depends on its ability to lose the heat produced in it by the current it carries and this in turn depends to some extent on the way the cable is installed. A cable clipped to a surface will more easily be able to dissipate heat than a similar cable which is installed with others in a conduit.

Cables must be protected from radiant heat, from moisture and from solid foreign bodies. They must be protected from corrosion and from impact or strain; for example, cables must be properly supported.

[Table 4A1] of [Appendix 4] gives a schedule of installation methods; for example, bare conductors and non-sheathed cables must not be clipped direct or fixed to cable ladder, tray or brackets. [Table 4A2] of [Appendix 4] lists one hundred and three standard methods of installation, each of them taken into account in the rating tables of the same Appendix. For example, two 2.5 mm^2 single core p.v.c. insulated non-armoured cables drawn into a steel conduit (installation method B) have a current rating of 24 A [Table 4.D1A]. A 2.5 mm^2 twin p.v.c. insulated and sheathed cable, which contains exactly the same conductors, has a current rating of 27 A [Table 4.D2A] when clipped directly to a non-metallic surface. Cables sheathed in p.v.c. must not be subjected to direct sunlight, because the ultra-violet component will leach out the plasticiser, causing the sheath to harden and crack. Cables must not be run in the same enclosure (e.g. trunking, pipe or ducting) as non-electrical services such as water, gas, air, etc. unless it has been established that the electrical system can suffer no harm as a result. If electrical and other services have metal sheaths and are touching, they must be bonded. Cables must not be run in positions where they may suffer or cause damage or interference with other systems. They should not, for example, be run alongside hot pipes or share a space with a hearing induction loop.

Cables must be protected from radiant heat, from moisture and from solid foreign bodies. They must be protected from corrosion and from impact or strain; for example, cables must be properly supported. Special precautions may need to be taken where cables or equipment are subject to ionising radiation. Where a wiring system penetrates a load bearing part of a building construction it must be ensured that the penetration will not adversely affect the integrity of the construction.

The build-up of dust on cables can act as thermal insulation. In some circumstances the dust may be flammable or even explosive. Cable runs must be designed to minimise dust accumulation and cables run on vertically mounted cable ladders rather than horizontal cable trays. When cables are run together, each sets up a magnetic field with a strength depending on the current carried. This field surrounds other cables, so that there is the situation of current-carrying conductors situated in a magnetic field. This will result in a force on the conductor, which is usually negligible under normal conditions but which can become very high indeed when heavy currents flow under fault conditions. All cables and conductors must be properly fixed or supported to prevent damage to them under these conditions.

4.3.4 Ambient temperature correction factors [Table 4B1 of Appendix 4]

The transfer of heat, whether by conduction, convection or radiation, depends on temperature difference - heat flows from hot to cold at a rate that depends on the temperature difference between them. Thus, a cable installed near the roof of a boiler house where the surrounding (ambient) temperature is very high will not dissipate heat so readily as one clipped to the wall of a cold wine cellar.

[Appendix 4] includes three tables giving correction factors to take account of the ability of a cable to shed heat due to the ambient temperature. Table 4B1 deals with air temperatures and is thus the Table likely to be of particular interest to the average electrician. Table 4B2 deals with buried cables, whilst 4B3 also concerns such cables installed within soils of unusual thermal resistivities. The Regulations use the symbol Ca to represent this correction factor. The tables assume that the ambient temperature is 30°C, and give a factor by which current rating is multiplied for other ambient temperatures.

For example, if a cable has a rating of 24 A and an ambient temperature correction factor of 0.77, the new current rating becomes 24 x 0.77 or 18.5 A. Different values are given for thermoplastic (PVC), thermosetting (rubber) and mineral insulation. The most useful of the correction factors are given in {Table 4.3}.

In [Table 4.3], '70°C m.i.' gives data for mineral insulated cables with sheaths covered in p.v.c. or LSF or open to touch, and '105°C m.i.' for mineral insulated cables with bare sheaths which cannot be touched and are not in contact with combustible material. The cable that is p.v.c. sheathed or can be touched must run cooler than if it is bare and not in contact with combustible material, and so has lower correction factors. Mineral insulated cables must have insulating sleeves in terminations with the same temperature rating as the seals used.

Where a cable is subjected to sunlight, it will not be able to lose heat so easily as one which is shaded. This is taken into account by adding 20°C to the ambient temperature for a cable which is unshaded.

Table 4.3 Correction factors to current rating for ambient temperature (Ca)
(from [Table 4B1] of BS 7671: 2008)

Ambient temperature (°C)	70°C thermoplastic p.v.c.	85°C thermosetting rubber	Type of insulation 70°C m.i	105°C m.i.
25	1.03	1.02	1.07	1.04
30	1.00	1.00	1.00	1.00
35	0.94	0.96	0.93	0.96
40	0.87	0.91	0.85	0.92
45	0.79	0.87	0.78	0.88
50	0.71	0.82	0.67	0.84
55	0.61	0.76	0.57	0.80

4.3.5 Cable grouping correction factors [Table 4C1 of Appendix 4]

If a number of cables is installed together and each is carrying current, they will all warm up. Those which are on the outside of the group will be able to transmit heat outwards, but will be restricted in losing heat inwards towards other warm cables. Cables 'buried' in others near the centre of the group may find it impossible to shed heat at all, and will rise further in temperature {Fig 4.9}.

Because of this, cables installed in groups with others (for example, if enclosed in a conduit or trunking) are allowed to carry less current than similar cables clipped to, or lying on, a solid surface that can dissipate heat more easily. If surface mounted cables are touching, the reduction in the current rating is, as would be expected, greater than if they are separated. {Figure 4.9} illustrates the difficulty of dissipating heat in a group of cables.

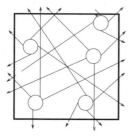 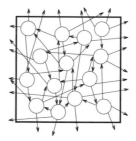

Fig 4.9 The need for the grouping correction factor C_g
a) widely spaced cables dissipate heat easily
b) closely packed cables cannot easily dissipate heat so their temperature rises

For example, if a certain cable has a basic current rating of 24 A and is installed in a trunking with six other circuits (note carefully, this is circuits and not cables), C_g has a value of 0.57 and the cable current rating becomes 24 x 0.57 or 13.7 A. The symbol C_g is used to represent the factor used for derating cables to allow for grouping. {Table 4.4} shows some of the more useful values of C_g.

The grouping factors are based on the assumption that all cables in a group are carrying rated current. If a cable is expected to carry no more than 30% of its grouped rated current, it can be ignored when calculating the group rating factor. For example, if there are four circuits in a group but one will be carrying less than 30% of its grouped rating, the group may be calculated on the basis of having only three circuits.

Table 4.4 Correction factors for groups of cables
(from [Table 4C1] of BS 7671: 2008)

Number of circuits	Correction factor Cg	
	Enclosed or clipped Touching	Clipped to non-metallic surface
2	0.80	0.85
3	0.70	0.79
4	0.65	0.75
5	0.60	0.73
6	0.57	0.72
7	0.54	0.72
8	0.52	0.71
9	0.50	0.70
12	0.45	0.70

4.3.6 Thermal insulation correction factors [523.9, Table 52.2]

The use of thermal insulation in buildings, in the forms of cavity wall filling, roof space blanketing, and so on, is now standard. Since the purpose of such materials is to limit the transfer of heat, they will clearly affect the ability of a cable to dissipate the heat build up within it when in contact with them. The Regulations state clearly that wherever possible cables should never be covered by heat insulating material, but this is clearly impractical in many installations. The cable rating tables of [Appendix 4] allow for the reduced heat loss for a cable that is enclosed in an insulating wall and is assumed to be in contact with the insulation on one side. In all other cases, the cable should be fixed in a position where it is unlikely to be completely covered by the insulation. Where this is not possible and a cable is buried in thermal insulation for 0.5 m (500 mm) or more, a rating factor (the symbol for the thermal insulation factor is Ci) of 0.5 is applied, which means that the current rating is halved.

Table 4.5 Derating factors (Ci) for cables up to 10 mm² in cross-sectional area buried in thermal insulation (from [Table 52.2] of BS 7671: 2008)

Length in insulation (mm)	Derating factor (Ci)
50	0.88
100	0.78
200	0.63
400	0.51
500 or more	0.50

If a cable is totally surrounded by thermal insulation for only a short length (for example, where a cable passes through an insulated wall), the heating effect on the insulation will not be so great because heat will be conducted from the short high-temperature length through the cable conductor. Clearly, the longer the length of cable enclosed in the insulation the greater will be the derating effect. {Table 4.5} shows the derating factors for lengths in insulation of up to 500 mm and applies to cables having cross-sectional area up to 10 mm².

Commonly-used cavity wall fillings, such as polystyrene sheets or granules, will have an adverse effect on p.v.c. sheathing, leeching out some of the plasticiser so that the p.v.c. becomes brittle. In such cases, an inert barrier must be provided to separate the cable from the thermal insulation. PVC cable in contact with bitumen may have some of its plasticiser removed: whilst this is unlikely to damage the cable, the bitumen will become fluid and may run.

4.3.7 When a number of correction factors applies

In some cases all the correction factors will need to be applied because there are parts of the cable which are subject to all of them. For example, if a mineral insulated cable with p.v.c. sheath protected by a circuit breaker and with a tabulated rated current of 34 A is run within the insulated ceiling of a boiler house with an ambient temperature of 45°C and forms part of a group of four circuits, derating will be applied as follows:

Actual current rating (I_z)
= tabulated current (I_t) x ambient temperature factor(C_a) x group factor (C_g) x thermal insulation factor (C_i)
= 34 x 0.78 x 0.65 x 0.5 A = **8.62 A**

In this case, the current rating is about one quarter of its tabulated value due to the application of correction factors. A reduction of this sort will only occur when all the correction factors apply at the same time. There are many cases where this is not so. If, for example, the cable above were clipped to the ceiling of the boiler house and not buried in thermal insulation, the thermal insulation factor would not apply.

Then, I_z = I_t x C_a x C_g = 34 x 0.78 x 0.65 A = **17.2 A**

The method is to calculate the overall factor for each set of cable conditions and then to use the lowest only. For example, if on the way to the boiler house the cable is buried in thermal insulation in the wall of a space where the temperature is only 20°C and runs on its own, not grouped with other circuits, only the correction factor for thermal insulation would apply. However, since the cable is then grouped with others, and is subject to a high ambient temperature, the factors are:

C_i = 0.5
C_a x C_g = 0.85 x 0.60 = 0.51

The two factors are almost the same, so either (but not both) can be applied. Had they been different, the smaller would have been used.

4.3.8 Protection by semi-enclosed (rewirable) fuses [Appendix 3, Appendix 4]

If the circuit concerned is protected by a semi-enclosed (rewirable) fuse the cable size will need to be larger to allow for the fact that such fuses are not so certain in operation as are cartridge fuses or circuit breakers. The fuse rating must never be greater than 0.725 times the current carrying capacity of the lowest-rated conductor protected.

In effect, this is the same as applying a correction factor of 0.725 to all circuits protected by semi-enclosed (rewirable) fuses.

4.3.9 Cable rating calculation [Appendix 4]

The following symbols are used when selecting cables:

I_z is the current carrying capacity of the cable in the situation where it is installed
I_t is the tabulated current for a single circuit at an ambient temperature of 30°C
I_b is the design current, the actual current to be carried by the cable
I_n is the rating of the protecting fuse or circuit breaker
I_2 is the operating current for the fuse or circuit breaker (the current at which the fuse blows or the circuit breaker opens)
C_a is the correction factor for ambient temperature
C_g is the correction factor for grouping
C_i is the correction factor for thermal insulation.

The correction factor for protection by a semi-enclosed (rewirable) fuse is not given a symbol but has a fixed value of 0.725.

Under all circumstances, the cable current carrying capacity must be equal to or greater than the circuit design current and the rating of the fuse or circuit breaker must be at least as big as the circuit design current. These requirements are common sense, because otherwise the cable would be overloaded or the protection would operate when the load is switched on. To ensure correct protection from overload, it is important that the protective device operating current (I_2) is not bigger than 1.45 times the current carrying capacity of the cable (I_z). Additionally, the rating of the fuse or circuit breaker (I_n) must not be greater than the cable current carrying capacity (I_z). It is important to appreciate that the operating current of a protective device is always larger than its rated value. In the case of a back-up fuse, which is not intended to provide overload protection, neither of these requirements applies.

To select a cable for a particular application, take the following steps: (note that to save time it may be better first to ensure that the expected cable for the required length of circuit will not result in the maximum permitted volt drop being exceeded {4.3.11}).

1 Calculate the expected (design) current in the circuit (I_b)
2 Choose the type and rating of protective device (fuse or circuit breaker) to be used (I_n)
3 Divide the protective device rated current by the ambient temperature correction factor (C_a) if ambient temperature differs from 30°C
4 Further divide by the grouping correction factor (C_g)
5 Divide again by the thermal insulation correction factor (C_i)
6 Divide by the semi-enclosed fuse factor of 0.725 where applicable
7 The result is the rated current of the cable required, which must be chosen from the appropriate tables {4.6 to 4.9}

Observe that one should divide by the correction factors, whilst in the previous subsection we were multiplying them. The difference is that here we start with the design current of the circuit and adjust it to take account of factors which will derate the cable. Thus, the current carrying capacity of the cable will be equal to or greater than the design current. In {4.3.7} we were calculating by how much the current carrying capacity was reduced due to application of correction factors. Volt drop and current carrying capacity of busbar trunking and powertrack systems is covered in [Appendix 8].

{Tables 4.6 to 4.9} give current ratings and volt drops for some of the more commonly used cables and sizes. The Tables assume that the conductors and the insulation are operating at their maximum rated temperatures. They are extracted from the Regulations Tables shown in square brackets e.g. [4D1A].The examples below will illustrate the calculations, but do not take account of volt drop requirements (see {4.3.11}).

Example 4.1

An immersion heater rated at 230 V, 3 kW is to be installed using twin with protective conductor p.v.c. insulated and sheathed cable. The circuit will be fed from a 16 A miniature circuit breaker type B, and will be run for much of its 14 m length in a roof space which is thermally insulated with glass fibre. The roof space temperature is expected to rise to 50°C in summer, and where it leaves the consumer unit and passes through a 50 mm insulation-filled cavity, the cable will be bunched with seven others. Calculate the cross-sectional area of the required cable.

First calculate the design current (I_b).

$$I_b = \frac{P}{U} = \frac{3000}{230} \text{ A} = 13.0 \text{ A}$$

The ambient temperature correction factor is found from {Table 4.3} to be 0.71. The group correction factor is found from {Table 4.4} as 0.52. (The circuit in question is bunched with seven others, making eight in all). The thermal insulation correction factor is already taken into account in the current rating table [4D2A ref. method A] and need not be further considered. This is because we can assume that the cable in the roof space is enclosed by the glass fibre. What we must consider is the point where the bunched cables pass through the insulated cavity. From {Table 4.5} we have a factor of 0.88.

The correction factors must now be considered to see if more than one of them applies to the same part of the cable. The only place where this happens is in the insulated cavity behind the consumer unit. Factors of 0.52 (C_g) and 0.88 (C_i) apply. The combined value of these (0.458), which is lower than the ambient temperature correction factor of 0.71, and will thus be the figure to be applied. Hence the required current rating is calculated:-

$$I_z = \frac{I_n}{C_g \times C_i} = \frac{16}{0.52 \times 0.88} \text{ A} = 35.0 \text{ A}$$

From {Table 4.7}, 6 mm² p.v.c. twin with protective conductor has a current rating of 32 A. This is not large enough, so 10 mm² with a current rating of 43 A is indicated. Not only would this add considerably to the costs, but would also result in difficulties due to terminating such a large cable in the accessories.

A more sensible option would be to look for a method of reducing the required cable size. For example, if the eight cables left the consumer unit in two bunches of four, this would result in a grouping factor of 0.65 (from {Table 4.4}). Before applying this, we must check that the combined grouping and thermal insulation factors (0.65 x 0.88 = 0.458) are still less than the ambient temperature factor of 0.71, which is the case.

This leads to a cable current rating of $\dfrac{16}{0.65 \times 0.88} \text{ A} = 28.0 \text{ A}$

This is well below the rating for 6 mm² of 32 A, so a cable of this size could be selected.

Table 4.6 Current ratings and volt drops for single-core thermoplastic
(p.v.c.) insulated cables (from [Tables 4D1A and 4D1B] of BS 7671: 2008)

Cross -sec area (mm²)	In conduit in thermal insulation (A)	In conduit in thermal insulation (A)	In conduit on wall (A)	In conduit on wall (A)	Clipped direct (A)	Clipped direct (A)	Volt drop (mV/ A/m)	Volt drop (mV/ A/m)
	2 cables	3 or 4 cables	2 cables	3 or 4 cables	2 cables	3 or 4 cables	2 cables	3 or 4 cables
1.0	11.0	10.5	13.5	12.0	15.5	14.0	44.0	38.0
1.5	14.5	13.5	17.5	15.5	20.0	18.0	29.0	25.0
2.5	20.0	18.0	24.0	21.0	27.0	25.0	18.0	15.0
4.0	26.0	24.0	32.0	28.0	37.0	33.0	11.0	9.5
6.0	34.0	31.0	41.0	36.0	47.0	43.0	7.3	6.4
10.0	46.0	42.0	57.0	50.0	65.0	59.0	4.4	3.8
16.0	61.0	56.0	76.0	68.0	87.0	79.0	2.8	2.4

Table 4.7 Current ratings and volt drops for multi-core thermoplastic
(p.v.c.) insulated cables (from [Tables 4D2A and 4D2B] of BS 7671:
2008). Note that the classification '3 or 4 core' refers to three-phase circuits and
not to applications such as two way switching.

Cross -sec area (mm²)	In wall in thermal insulation (A)	In wall in thermal insulation (A)	In conduit on wall (A)	In conduit on wall (A)	Clipped direct (A)	Clipped direct (A)	Volt drop (mV/) A/m	Volt drop (mV/) A/m
	2 core	3 or 4 core	2 core	3 or 4 core	2 core	3 or 4 core	2 core	3 or 4 core
1.0	11.0	10.0	13.0	11.5	15.0	13.5	44.0	38.0
1.5	14.0	13.0	16.5	15.0	19.5	17.5	29.0	25.0
2.5	18.5	17.5	23.0	20.0	27.0	24.0	18.0	15.0
4.0	25.0	23.0	30.0	27.0	36.0	32.0	11.0	9.5
6.0	32.0	29.0	38.0	34.0	46.0	41.0	7.3	6.4
10.0	43.0	39.0	52.0	46.0	63.0	57.0	4.4	3.8
16.0	57.0	52.0	69.0	62.0	85.0	76.0	2.8	2.4

Table 4.7A Current carrying capacities and volt drops for 70°C thermo-
plastic (pvc) insulated and sheathed flat cable with protective
conductors (from [Tables 4D5] of BS 7671: 2008).

Conductor c.s.a. (mm²)	In conduit in insulated wall (A)	Directly insulated wall (A)	Clipped direct (A)	Volt drop per A per m
1.0	11.5	12.0	16.0	44.0
1.5	14.5	15.0	20.0	29.0
2.5	20.0	21.0	27.0	18.0
4.0	26.0	27.0	37.0	11.0
6.0	32.0	35.0	47.0	7.3
10.0	44.0	47.0	64.0	4.4
16.0	57.0	63.0	85.0	2.8

Example 4.2

The same installation as in Example 4.1 is proposed. To attempt to make the cable
size smaller, the run in the roof space is to be kept clear of the glass fibre insulation.
Does this make a difference to the selected cable size?

There is no correction factor for the presence of the glass fibre, so the calculation of I_z will be exactly thee same as for Example 4.1 at 35.0 A. This time reference method 1 (clipped direct) will apply to the current rating {Table 4.7}. For a two-core cable, 4mm² has a current rating of 36 A so this will be the selected size.

It is of interest to notice how quite a minor change in the method of installation, in this case clipping the cable to joists or battens clear of the glass fibre, has reduced the acceptable cable size.

Table 4.8   Current ratings of mineral insulated cables clipped direct
(from [Tables 4G1A and 4G2A] of BS 7671: 2008)

Cross-sec. area (mm²)		p.v.c. sheath 2 x single or twin (A)	p.v.c. sheath 3 core (A)	p.v.c. sheath 3 x single (A)	bare sheath 2 x single or twin (A)	bare sheath 3 x single (A)
1.0	(500V)	18.5	15.0	17.0	22.0	21.0
1.5	(500V)	23.0	19.0	21.0	28.0	27.0
2.5	(500V)	31.0	26.0	29.0	38.0	36.0
4.0	(500V)	40.0	35.0	38.0	51.0	47.0
1.0	(750V)	19.5	16.0	18.0	24.0	24.0
1.5	(750V)	25.0	21.0	23.0	31.0	30.0
2.5	(750V)	34.0	28.0	31.0	42.0	41.0
4.0	(750V)	45.0	37.0	41.0	55.0	53.0
6.0	(750V)	57.0	48.0	52.0	70.0	67.0
10.0	(750V)	77.0	65.0	70.0	96.0	91.0
16.0	(750V)	102.0	86.0	92.0	127.0	119.0

Note that in {Tables 4.8 and 4.9} 'P.V.C. sheath' means bare and exposed to touch or having an overall covering of p.v.c. or LSF and 'Bare' means bare and neither exposed to touch nor in contact with combustible materials.

Table 4.9   Volt drops for mineral insulated cables
(from [Tables 4G1B and 4G2B] of BS 7671: 2008)

Cross-sec. area (mm²)	Single-phase p.v.c. sheath (mV/A/m)	Single-phase bare (mV/A/m)	Three-phase p.v.c. sheath (mV/A/m)	Three-phase bare (mV/A/m)
1.0	42.0	47.0	36.0	40.0
1.5	28.0	31.0	24.0	27.0
2.5	17.0	19.0	14.0	16.0
4.0	10.0	12.0	9.1	10.0
6.0	7.0	7.8	6.0	6.8
10.0	4.2	4.7	3.6	4.1
16.0	2.6	3.0	2.3	2.6

Example 4.3

Assume that the immersion heater indicated in the two previous examples is to be installed, but this time with the protection of a 15 A rewirable (semi-enclosed) fuse. Calculate the correct cable size for each of the alternatives, that is where firstly the cable is in contact with glass fibre insulation, and secondly where it is held clear of it.

This time the value of the acceptable current carrying capacity I_z will be different because of the need to include a factor for the rewirable fuse as well as the new ambient temperature and grouping factors for the rewirable fuse from {Tables 4.3 and 4.4}.

$$I_z = \frac{I_n}{C_g \times C_a \times 0.725} = \frac{15}{0.52 \times 0.89 \times 0.725} = 44.7 \text{ A}$$

In this case, the cable is in contact with the glass fibre, so the first column of {Table 4.7} of current ratings will apply. The resulting cable size is 16 mm² which has a current rating of 57 A. This cable size is not acceptable on the grounds of high cost and because the conductors are likely to be too large to enter the connection tunnels of the immersion heater and its associated switch. If the cables leaving the consumer unit are re-arranged in two groups of four, this will reduce the grouping factor to 0.65, so that the newly calculated value of I_z is 35.8 A. This means using 10 mm² cable with a current rating of 43 A (from {Table 4.7}), since 6 mm² cable is shown to have a current rating in these circumstances of only 32 A. By further rearranging the cables leaving the consumer unit to be part of a group of only two, C_g is increased to 0.8, which reduces I_z to 29.1 A which enables selection of a 6 mm² cable.

Should it be possible to bring the immersion heater cable out of the consumer unit on its own, no grouping factor would apply and I_z would fall to 23.2 A, allowing a 4 mm² cable to be selected.

Where the cable is not in contact with glass fibre there will be no need to repeat the calculation of I_z, which still has a value of 29.1 A provided that it is possible to group the immersion heater cable with only one other where it leaves the consumer unit. This time we use the 'clipped direct)' column of the current rating {Table 4.7}, which shows that 4 mm² cable with a current rating of 36 A will be satisfactory.

Examples 4.1, 4.2 and 4.3 show clearly how forward planning will enable a more economical and practical cable size to be used than would appear necessary at first. It is, of course, important that the design calculations are recorded and retained in the installation manual.

Example 4.4

A 400 V 50 Hz three-phase motor with an output of 7.5 kW, power factor 0.8 and efficiency 85% is the be wired using 500 V light duty three-core mineral insulated p.v.c. sheathed cable. The length of run from the HBC protecting fuses is 20 m, and for about half this run the cable is clipped to wall surfaces. For the remainder it shares a cable tray, touching two similar cables across the top of a boiler room where the ambient temperature is 50°C. Calculate the rating and size of the correct cable.

The first step is to calculate the line current of the motor

$$\text{Input} = \frac{\text{output}}{\text{efficiency}} = \frac{7.5 \times 100}{85} \text{ kW} = 8.82 \text{ kW}$$

$$\text{Line current } I_b = \frac{P}{\sqrt{3} \times U_L \times \cos\phi} = \frac{8.82 \times 10^3}{\sqrt{3} \times 400 \times 0.8} \text{ A} = 15.9 \text{ A}$$

We must now select a suitable fuse. {Fig 3.15} for BS 88 fuses shows the 16 A size to be the most suitable. Part of the run is subject to an ambient temperature of 50°C, where the cable is also part of a group of three, so the appropriate correction factors must be applied from {Tables 4.3 and 4.4}.

$$I_z = \frac{I_n}{C_g \times C_a} = \frac{16}{0.70 \times 0.67} \text{ A} = 34.1 \text{ A}$$

Note that the grouping factor of 0.70 has been selected because where the cable is grouped it is clipped to a metallic cable tray, and not to a non-metallic surface. Next the cable must be chosen from {Table 4.8}. Whilst the current rating would be 15.9 A if all of the cable run were clipped to the wall, part of the run is subject to the two correction factors, so a rating of 34.1 A must be used. For the clipped section of the cable (15.9 A), reference method 1 could be used which gives a size of 1.5 mm² (current rating 21.0 A). However, since part of the cable is on the tray (method 3) the correct size for 34.1 A will be 4.0 mm², with a rating of 37 A.

4.3.10 Special formulas for grouping factor calculation [App. 4]
In some cases the conductor sizes where cables are grouped and determined by the methods shown in {4.3.9} can be reduced by applying some rather complex formulas given in [Appendix 4]. Whilst it is true that in many cases the use of these formulas will show the installer that it is safe for him to use a smaller cable than he would have needed by simple application of correction factors, this is by no means always the case. There are many cases where their application will make no difference at all. Since this book is for the electrician, rather than for the designer, the rather complicated mathematics will be omitted.

4.3.11 Cable volt drop [525, Appendix 4, Appendix 8, Appendix 12]
All cables have resistance, and when current flows in them this results in a volt drop. Hence, the voltage at the load is lower than the supply voltage by the amount of this volt drop. The volt drop may be calculated using the basic Ohm's law formula

$$U = I \times R$$
where U is the cable volt drop (V)
I is the circuit current (A), and
R is the circuit resistance (Ω)

Unfortunately, this simple formula is seldom of use in this case, because the cable resistance under load conditions is not easy to calculate. [525] indicates that the voltage at any load must never fall so low as to impair the safe working of that load, or fall below the level indicated by the relevant British Standard where one applies.

[Appendix 12] states that these requirements will be met if the voltage drop does not exceed 3% of the declared supply voltage for a lighting system, or 5% for other applications. This is at variance with the Electrical Safety, Quality and Continuity (ESCQ) Regulations, that require an upper volt drop limit of 4% for all systems.

It should be borne in mind that European Agreement HD 472 S2 allows the declared supply voltage to vary by ±10%. Assuming that the supply voltage of 230 V is 10% low, and allowing a 5% volt drop, this gives permissible load voltages of 195.5 V for a single-phase 230V supply, or 340.0 V (line) for a 400 V three-phase supply.

To calculate the volt drop for a particular cable we use {Tables 4.6, 4.7, 4.7A and 4.9} which are extracted from [Appendix 4] tables. Each current rating table has an associated volt drop column or table. For example, multicore sheathed non-armoured p.v.c. insulated cables are covered by {Table 4.7} for current ratings, and volt drops. The exception in the Regulations to this layout is for mineral insulated cables where there are separate volt drop tables for single- and three-phase operation, which are combined here as {Table 4.9}. Volt drop and current carrying capacity of busbar trunking and powertrack systems is covered in [Appendix 8].

Each cable rating in the Tables of [Appendix 4] has a corresponding volt drop figure in millivolts per ampere per metre of run (mV/A/m). Strictly this should be mV/Am, but here we shall follow the pattern adopted by BS 7671: 2008. To calculate the cable volt drop:

1 take the value from the volt drop table (mV/A/m)
2 multiply by the actual current in the cable (NOT the current rating)
3 multiply by the length of run in metres
4 divide the result by one thousand (to convert millivolts to volts).

For example, if a 4 mm² p.v.c. sheathed circuit feeds a 6 kW shower and has a run length of 16 m, we can find the volt drop thus:

From {Table 4.7}, the volt drop figure for 4 mm² two-core cable is 11 mV/A/m.

Cable current is calculated from $I = \dfrac{P}{U} = \dfrac{6000}{230}$ A $= 26.1$ A

Volt drop is then $\dfrac{11 \times 26.1 \times 16}{1000}$ V $= 4.6$ V

Since the permissible volt drop is 5% of 230 V, which is 11.5 V, the cable in question meets volt drop requirements. The following examples will make the method clear.

Example 4.5

Calculate the volt drop for the case of Example 4.1. What maximum length of cable would allow the installation to comply with the volt drop regulations?

The table concerned here is {4.7}, which shows a figure of 7.3 mV/A/m for 6 mm² twin with protective conductor pvc insulated and sheathed cable. The actual circuit current is 13.0 A, and the length of run is 14 m.

Volt drop $= \dfrac{7.3 \times 13.0 \times 14}{1000}$ V $= 1.33$ V

Maximum permissible volt drop is 5% of 230 V $= \dfrac{5 \times 230}{100}$ V $= 11.5$ V

If a 14 m run gives a volt drop of 1.33 V, the length of run for an 11.5 V drop will be:

$\dfrac{11.5 \times 14}{1.33}$ m $= 121$ m

Example 4.6

Calculate the volt drop for the case of {Example 4.2}. What maximum length of cable would allow the installation to comply with the volt drop regulations?

The Table concerned here is {4.7} which shows a volt drop figure for 4.0 mm² cable of 11 mV/A/m, with the current and the length of run remaining at 13.0 A and 14 m respectively.

Volt drop $= \dfrac{11 \times 13.0 \times 14}{1000}$ V $= 2.0$ V

Maximum possible volt drop is 5% of 230 V $= \dfrac{5 \times 230}{100}$ V $= 11.5$ V

If a 14 m run gives a volt drop of 2.0 V, the length of run for a 11.5 V drop will be

$\dfrac{11.5 \times 14}{2.0}$ m $= 80.5$ m

Example 4.7

Calculate the volt drop for the cases of {Example 4.3} for each of the alternative installations. What maximum length of cable would allow the installation to comply with the volt drop regulations in each case?

In neither case is there any change in cable sizes, the selected cables being 6 mm² in the first case and 4 mm² in the second. Solutions are thus the same as those in {Examples 4.5 and 4.6} respectively.

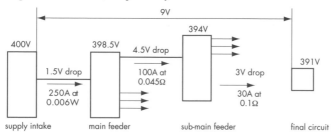

Fig 4.10 Total volt drop in large installations

Example 4.8

Calculate the volt drop and maximum length of run for the motor circuit of {Example 4.4}.

This time we have a mineral insulated p.v.c. sheathed cable, so volt drop figures will come from {Table 4.9}. This shows 9.1 mV/A/m for the 4 mm² cable selected, which must be used with the circuit current of 15.9 A and the length of run which is 20 m.

Volt drop will be $\dfrac{9.1 \times 15.9 \times 20}{1000}$ V = 2.89 V

Permissible maximum volt drop is 5% of 400 V or $\dfrac{5 \times 400}{100}$ V = 20.0 V

Maximum length of run for this circuit with the same cable size and type will be

$\dfrac{20.0 \times 20}{2.89}$ m = 138 m

The 'length of run' calculations carried out in these examples are often useful to the electrician when installing equipment at greater distances from the mains position. It is important to appreciate that the allowable volt drop of 3% (lighting) or 5% (other) of the supply voltage applies to the whole of an installation. If an installation has mains, submains and final circuits, for instance, the volt drop in each must be calculated and added to give the total volt drop as indicated in {Fig 4.10}.

All of our work in this sub-section so far has assumed that cable resistance is the only factor responsible for volt drop. In fact, larger cables have significant self inductance as well as resistance. As we shall see in Chapter 5 there is also an effect called impedance which is made up of resistance and inductive reactance (see {Fig 5.8(a)}).

Inductive reactance $X_L = 2\pi f_L$

where X_L = inductive reactance in ohms (Ω)

π = the mathematical constant 3.142

f = the system frequency in hertz (Hz)

L = circuit self inductance in henrys (H)

It is clear that inductive reactance increases with frequency, and for this reason the volt drop tables apply only to systems with a frequency lying between 49 Hz and 51 Hz.

For small cables, the self inductance is such that the inductive reactance, is small compared with the resistance. Only with cables of cross-sectional area 25 mm^2 and greater need reactance be considered. Since cables as large as this are seldom used on work that has not been designed by a qualified engineer, the subject of reactive volt drop will not be further considered here.

If the actual current carried by the cable (the design current) is less than the rated value, the cable will not become as warm as the calculations used to produce the volt drop tables have assumed. The Regulations include (in [Appendix 4]) a very complicated formula to be applied to cables of cross-sectional area 16 mm^2 and less which may show that the actual volt drop is less than that obtained from the tables. This possibility is again seldom of interest to the electrician, and is not considered here.

4.3.12 Harmonic currents and neutral conductors [Appendix 8, Appendix 11]

A perfectly balanced three-phase system (one with all three phase loads identical in all respects) has no neutral current and thus has no need of a neutral conductor. This is often so with motors, which are fed through three core cables in most cases. Many three-phase loads are made up of single-phase loads, each connected between one line and neutral. It is not likely in such cases that the loads will be identical, so the neutral will carry the out-of-balance current of the system. The greater the degree of imbalance, the larger the neutral current. Where discharge lighting is fed through a three phase supply, or where the phase currents of such a supply have harmonic current content of more than 10%, the neutral conductor must be no smaller than the line conductors.

Some three-phase four-core cables have a neutral of reduced cross-section on the assumption that there will be some degree of balance. Such a cable must not be used unless the installer is certain that severe out-of-balance conditions will never occur. Similar action must be taken with a three-phase circuit wired in single-core cables. A reduced neutral conductor may only be used where out-of-balance currents will be very small compared to the line currents.

A problem is likely to occur in systems which generate significant third harmonic currents. Devices such as discharge lamp ballasts, switched mode power supplies (often for computers) and transformers on low load distort the current waveform. Thus, currents at three times normal frequency (third harmonics) are produced, which do not cancel at the star point of a three-phase system as do basic frequency currents, but add up, so that the neutral carries very heavy third harmonic currents. For this reason, it is important not to reduce the cross-sectional area of a neutral used to feed discharge lamps (including fluorescent lamps).

In some cases the neutral current may be considerably larger than the phase currents. Where the load concerned is fed through a multi-core cable, it may be prudent to use five-core (or even six-core) cables, so that two (or three) conductors may be used in parallel for the neutral.

In some cases where the neutral current is reasonably expected to be greater than the phase current, it may be necessary to insert overload protection in a neutral conductor. Such protection must be arranged to open all phase conductors on operation, but not the neutral. This clearly indicates the use of a special circuit breaker.

It is very important that the neutral of each circuit is kept quite separate from those of other circuits. Good practice suggests that the separate circuit neutrals should be connected in the same order at the neutral block as the corresponding phase conductors at the fuses or circuit breakers.

4.3.13 Low smoke-emitting cables [527]

Normal p.v.c. insulation emits dense smoke and corrosive gases when burning. If cables are to be run in areas of public access, such as schools, supermarkets, hospitals, etc, the designer should consider the use of special LSF cables, such as those with thermo-setting or elastomeric insulation, which do not cause such problems in the event of fire. This action is most likely to be necessary in areas expected to be crowded, along fire escape routes, and where equipment is likely to suffer damage due to corrosive fumes.

4.3.14 The effects of animals, insects and plants (flora and fauna) [522.9, 522.10]

Cables may be subject to damage by animals and plants as well as from their environment. Rodents in particular seem to have a taste for some types of cable sheathing and can gnaw through sheath and insulation to expose the conductors. Cables impregnated with repellant chemicals are not often effective and may also fall foul of the Health and Safety Regulations. Rodents build nests, often of flammable materials, leading to a fire hazard. Care should be taken to avoid cable installation along possible vermin runs, but where this cannot be avoided, steel conduit may be the answer.

Mechanical damage to wiring systems by larger animals such as cattle and horses can often be prevented by careful siting of cable runs and outlets. Attention must also be given to the fact that waste products from animals may be corrosive. Access by insects is difficult to prevent, but vent holes can be sealed with breathers. Damage by plants is a possible hazard, the effect of tree roots on small lighting columns being an obvious problem area.

4.3.15 Maximum conductor operating temperatures [523.1]

The operating temperature of a cable conductor must be limited because of the damaging effect, not on the conductor itself, but on the insulation that is in contact with it. Where the conductor operating temperature exceeds 70°C there is a danger that the equipment to which the conductor is connected may be damaged by conducted heat, and cable size may need to be increased to limit the conductor temperature. Maximum conductor operating temperatures are those shown in Table 5.8, which also gives k values.

4.4 CABLE SUPPORTS, JOINTS AND TERMINATIONS

4.4.1 Cable supports and protection [522.8, 526]

Cables must be fixed securely at intervals that are close enough to ensure that there will be no excessive strain on the cable or on its joints and terminations, and to prevent cable loops appearing which could lead to mechanical damage. {Table 4.10} indicates minimum acceptable spacings of fixings for some common types of cables.

Where cable runs are neither vertical nor horizontal, the spacing depends on the angle as shown in {Fig 4.11}. Where a cable is flat in cross-section as in the case of a p.v.c. insulated and sheathed type, the overall diameter is taken as the major axis as shown in {Fig 4.12}.

Table 4.10 Maximum spacing for cable supports

Overall cable diameter (mm)	p.v.c. sheathed horizontal (mm)	vertical (mm)	Mineral insulated horizontal (mm)	vertical (mm)
up to 9	250	400	600	800
10 to 15	300	400	900	1200
16 to 20	350	450	1500	2000
21 to 40	400	550	2000	3000

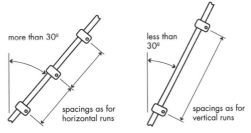

Fig 4.11 Spacing of support clips on angled runs

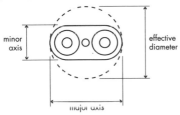

Fig 4.12 Effective diameter of a flat cable

Not only must hidden cables be protected from damage, but holes and notches to accommodate them must not weaken the floor in which they are cut. Holes must not be cut in roof rafters. Thus, where cables are run beneath boarded floors, they must pass through holes drilled in the joists which are at least 50 mm below the top surface of the joist. This is to prevent accidental damage due to nails being driven into the joists. The hole diameters must not exceed one quarter of the depth of the joist and they must be drilled at the joist centre (the neutral axis). Hole centres must be at least three diameters apart, and the holes must only be drilled in a zone which extends 25% to 40% of the beam length from both ends.

An alternative is to protect the cable in steel conduit. It is not practicable to thread rigid conduit through holes in the joists, so the steel conduit may be laid in slots cut in the upper or lower edges as shown in {Fig 4.13}. The depth of the slot must be no greater than one eighth of the joist depth and notches must be in a zone extending from 7% to 25% of the beam length from both ends. In practical terms, this means that for a joist 5m long which is supported at both ends, the notches must not be less than 35 cm or more than 1.25 m from either end. Notches must not be cut in roof rafters. These rules apply to dwelling houses which are no more than three storeys high.

It is common practice to conceal cables in chases cut into the wall. To ensure that the safety of the wall is not affected:

1 Vertical chases must be no deeper than one third of the thickness of the wall or of the inner leaf of a cavity wall.

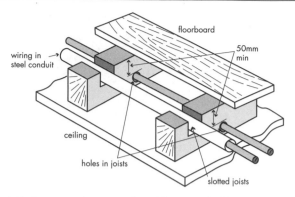

Fig 4.13 Support and protection for cables run under floors

2 Horizontal chases must be no deeper than one sixth of the thickness of the wall or of the inner leaf of a cavity wall.
3 The stability of the wall should not be affected by chases, a particular problem where hollow block walls are used.

Where cable runs are concealed behind plaster they must be installed in 'acceptable zones' which are intended to reduce the danger to the cables and to people who drill holes or knock nails into walls. Cable runs must only follow paths that are horizontal or vertical from an outlet, or be within 150 mm of the top (but not the bottom) of the wall, or within 150 mm of the angle formed by two adjoining walls. Where a cable run has to be diagonal, it must be protected by being enclosed in steel conduit, or must be a cable with an earthed metal sheath (such as mineral insulated cable), or an insulated concentric cable. In this latter case, the phase conductor will be surrounded by the neutral, so that if a nail or a screw penetrates the cable it will be impossible for it to become live. The possible zones are shown in {Fig 4.14}. Where these requirements cannot be met, the circuit(s) concerned must be protected by an RCD rated at no more than 30 mA.

The internal partition walls of some modern buildings are very thin, and where cables complying with the requirements above are within 50 mm of the surface on the other side, they will require protection.

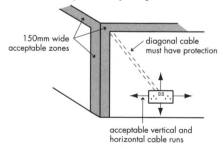

Fig 4.14 Acceptable installation zones for concealed cables. The diagonal cable must be enclosed in earthed metal or be protected by a 30 mA RCD,

There are cases where cables are enclosed in long vertical runs of trunking or conduit. The weight of the cable run, which effectively is hanging onto the top support, can easily cause damage by compression of the insulation where it is pulled against the support. In trunking there must be effective supports no more than 5 m apart, examples of which are shown in Fig 4.15, whilst for conduit the run must be provided with adaptable boxes at similar intervals which can accommodate the necessary supports.

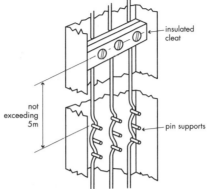

Fig 4.15 Support for vertical cables in trunking

The top of a vertical conduit or trunking run must have a rounded support to reduce compression of insulation. The diameters required will be the same as those for cable bends given in {4.4.2}. Support for overhead conductors is considered in {8.14}.

4.4.2 Cable bends [522.8]

If an insulated cable is bent too sharply, the insulation and sheath on the inside of the bend will be compressed, whilst that on the outside will be stretched. This can result in damage to the cable as shown in {Fig 4.16}. The bending factors that must be used to assess the minimum acceptable bending radius, values for common cables being given in {Table 4.11}. The factor shown in the table is that by which the overall cable diameter {Fig 4.12} must be multiplied to give the minimum inside radius of the bend. For example, 2.5 mm² twin with protective conductor sheathed cable has a cross-section 9.7 mm x 5.4 mm. Since the Table shows a factor of 3 for this size, the minimum inside radius of any bend must be 3 x 9.7 = 29.1 mm.

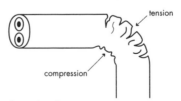

Fig 4.16 Damage to cable insulation due to bending

Table 4.11 Bending Factors for common cables

Type of insulation	Overall diameter	Bending factor
p.v.c.	up to 10 mm	3 (2)
p.v.c.	10 mm to 25 mm	4 (3)
p.v.c.	over 25 mm	6
mineral	any	6*

The figures in brackets apply to unsheathed single-core stranded p.v.c. cables when installed in conduit, trunking or ducting.
* Mineral insulated cables may be bent at a minimum radius of three times cable diameter provided that they will only be bent once. This is because the copper sheath will 'work harden' when bent and is likely to crack if straightened and bent again.

4.4.3 Joints and terminations [526]

The normal installation has many joints, and it follows that these must all remain safe and effective throughout the life of the system. With this in mind, regulations on joints include the following:

1 All joints must be durable, adequate for their purpose, and mechanically strong.

2 They must be constructed to take account of the conductor material and insulation, as well as temperature: eg, a soldered joint must not be used where the temperature may cause the solder to melt or to weaken. Very large expansion forces are not uncommon in terminal boxes situated at the end of straight runs of large cables when subjected to overload or to fault currents.

3 All joints and connections must be made in an enclosure complying with the appropriate British Standard.

4 Where sheathed cables are used, the sheath must be continuous into the joint enclosure {Figure 4.17}.

5 All joints must be accessible for inspection and testing unless they are buried in compound or encapsulated, are between the cold tail and element of a heater such as a pipe tracer or underfloor heating system, or are made by soldering, welding, brazing or compression.

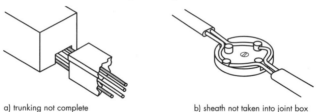

a) trunking not complete   b) sheath not taken into joint box

Fig 4.17 Failure to enclose non-sheathed cables

4.5 CABLE ENCLOSURES

4.5.1 Plastic and metal conduits [521.6, 522.3, 522.8]

A system of conduits into which unsheathed cables can be drawn has long been a standard method for electrical installations. The Regulations applying to conduit systems may be summarised as follows:

1 All conduits and fittings must comply with the relevant British Standards.

2 Plastic conduits must not be used where the ambient temperature or the temperature of the enclosed cables will exceed 60°C. Cables with thermosetting insulation are permitted to run very hot, and must be suitably downrated when installed in plastic conduit. To prevent the spread of fire, plastic conduits (and plastic trunking) must comply with ignitability characteristic P of BS 476 Part 5.

3 Conduit systems must be designed and erected so as to exclude moisture, dust and dirt. This means that they must be completely closed, with box lids fitted. To ensure that condensed moisture does not accumulate, small drainage holes must be provided at the lowest parts of the system.

4 Proper precautions must be taken against the effects of corrosion (see {4.2.5}), as well as against the effects of flora (plant growths) and fauna (animals). Protection from rusting of steel conduit involves the use of galvanised (zinc coated) tubing, and against electrolytic corrosion the prevention of contact between dissimilar metals eg steel and aluminium. Any additional protective

conductor must be run inside the conduit or its reactance is likely to be so high that it becomes useless if intended to reduce fault loop impedance.

5 A conduit system must be completely erected before cables are drawn in. It must be free of burrs or other defects which could damage cables whilst being inserted.

6 The bends in the system must be such that the cables drawn in will comply with the minimum bending radius requirements {4.4.2}.

7 The conduit must be installed so that fire cannot spread through it, or through holes cut in floors or walls to allow it to pass. This subject of fire spread will be considered in greater detail in {4.5.2}. All conduits must be non-flame propagating to BS EN 50086.

8 Allowance must be made, in the form of expansion loops, for the thermal expansion of long runs of metal or plastic conduit. Remember that plastic expands and contracts more than steel.

9 Use flexible joints when crossing building expansion points

Table 4.12 Maximum spacing of supports for conduits

Conduit diameter (mm)	Rigid metal (m) Horizontal	Vertical	Rigid insulating (m) Horizontal	Vertical
up to 16	0.75	1.0	0.75	1.0
16 to 25	1.75	2.0	1.5	1.75
25 to 40	2.0	2.25	1.75	2.0
over 40	2.25	2.5	2.0	2.0

4.5.2 Ducting and trunking [521.6, 521.10, 522.3, 522.8]

Metal and plastic trunkings are very widely used in electrical installations. They must be manufactured to comply with the relevant British Standards, and must be installed so as to ensure that they will not be damaged by water or by corrosion (see {4.2.5}).

Table 4.13 Support spacings for trunking

Maximum distances between supports are in metres.

Typical trunking size (mm)	Metal Horizontal	Vertical	Insulating Horizontal	Vertical
up to 25 x 25	0.75	1.0	0.5	0.5
up to 50 x 25	1.25	1.5	0.5	0.5
up to 50 x 50	1.75	2.0	1.25	1.25
up to 100 x 50	3.0	3.0	1.75	2.0

The data shown in {Tables 4.12 and 4.13} are not included in BS 7671: 2008, but are taken from Guidance Note 1, Selection and Erection, Appendix G.

If it is considered necessary to provide an additional protective conductor in parallel with steel trunking, it must be run inside the trunking or the presence of steel between the live and protective cables will often result in the reactance of the protective cable being so high that it will have an effect on fault loop impedance. Trunking must be supported as indicated in {Table 4.13}. The table does not apply to special lighting trunking which is provided with strengthened couplers. Where crossing a building expansion joint a suitable flexible joint should be included.

Where trunking or conduit passes through walls or floors the hole cut must be made good after the first fix on the construction site to give the partition the same degree of fire protection it had before the hole was cut. Holes should be kept as few and as small as possible. The fire-stopping material may need special support where it is possible that flammable materials will release it in the event of fire. Since it is possible for fire to spread through the interior of the trunking or conduit, fire barriers must be inserted as shown in {Fig 4.18}. An exception is conduit or trunking with a cross-sectional area of less than 710 mm², so that conduits up to 32 mm in diameter and trunking up to 25 mm x 25 mm need not be protected with fire barriers. During installation, temporary fire barriers must be provided so that the integrity of the fire prevention system is always maintained. All trunking must be non-flame propagating to BS EN 50085.

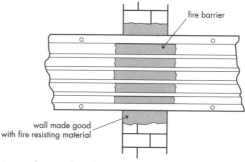

Fig 4.18 Provision of fire barriers in ducts and trunking

Since trunking will not be solidly packed with cables (see {4.5.3}) there will be room for air movement. A very long vertical trunking run may thus become extremely hot at the top as air heated by the cables rises; this must be prevented by barriers as shown in {Fig 4.19}. In many cases the trunking will pass through floors as it rises, and the fire stop barriers needed will also act as barriers to rising hot air.

Lighting trunking is being used to a greater extent than previously. In many cases, it includes copper conducting bars so that luminaires can be plugged in at any point, especially useful for display lighting. The considerably improved life, efficiency and colour rendering propties of extra-low voltage tungsten halogen lamps has led to their increasing use, often fed by lighting trunking. It is important here to remember that whilst the voltage of a 12 V lamp is only about one twentieth of normal mains potential, the current for the same power inputs will be more than twenty times greater. Thus, a trunking feeding six 50 W 12 V lamps will need to be rated at 25 A.

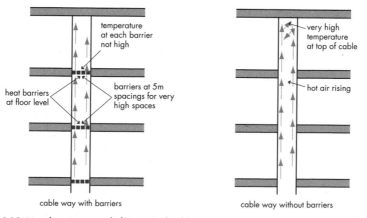

Fig 4.19 Heat barriers provided in vertical cable ways

4.5.3. Cable capacity of conduits and trunking [521.6, 528]

Not only must it be possible to draw cables into completed conduit and trunking systems, but neither the cables nor their enclosures must be damaged in the process. If too many cables are packed into the space available, there will be a greater increase in temperature during operation than if they were given more space. It is important to appreciate that grouping factors (see {4.3.5}) still apply to cables enclosed in conduit or trunking. To calculate the number of cables that may be drawn into a conduit or trunking, we make use of four tables ({Tables 4.14 to 4.17}). For situations not covered by these tables, the requirement is that a space factor of 45% must not be exceeded. This means that not more than 45% of the space within the conduit or trunking must be occupied by cables, and involves calculating the cross-sectional area of each cable, including its insulation, for which the outside diameter must be known. The cable factors for cables with thermosetting insulation are higher than those for pvc insulation when the cables are installed in trunking, but the two are the same when drawn into conduit (see {Table 4.14}). The data in {Tables 4.14 to 4.17} are not included in BS 7671: 2008, but are taken from Guidance Note 1, Selection and Erection, Appendix A.

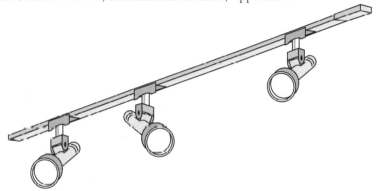

Fig 4.20 Low voltage luminaire on lighting trunking

The figures in {Table 4.14} may be high when applied to some types of plastic trunking due to the large size of the internal lid fixing clips. To use the tables {4.14 to 4.17}, the cable factors for all the conductors must be added. The conduit or trunking selected must have a factor (otherwise called 'term') at least as large as this number.

Table 4.14 Cable factors (terms) for conduit and trunking

Type of conductor	Conductor c.s.a. (mm²)	Factor for conduit	Factor for trunking pvc insulation	Factor for trunking thermosetting insulation
solid	1.0	16	3.6	3.8
solid	1.5	22	8.0	8.6
stranded	1.5	22	8.6	9.1
solid	2.5	30	11.9	11.9
stranded	2.5	30	12.6	13.9
stranded	4.0	43	16.6	18.1
stranded	6.0	58	21.2	22.9
stranded	10.0	105	35.3	36.3
stranded	16.0	145	47.8	50.3
stranded	25.0	217	73.9	75.4

Table 4.15 Cable factors (terms) for short straight runs up to 3 m

Type of conductor	Conductor c.s.a. (mm²)	Cable factor
solid	1.0	22
solid	1.5	27
solid	2.5	39
stranded	1.5	31
stranded	2.5	43
stranded	4.0	58
stranded	6.0	88
stranded	10.0	146

Table 4.16 Conduit factors (terms)

Length of run between boxes (m)

	1	2	3	4	5	6	8	10
conduit, straight								
16mm	290	290	290	177	171	167	158	150
20mm	460	460	460	286	278	270	256	244
25mm	800	800	800	514	500	487	463	442
32mm	1400	1400	1400	900	878	857	818	783
conduit, one bend								
16mm	188	177	167	158	150	143	130	120
20mm	303	286	270	256	244	233	213	196
25mm	543	514	487	463	442	422	388	358
32mm	947	900	857	818	783	750	692	643
conduit, two bends								
16mm	177	158	143	130	120	111	97	86
20mm	286	256	233	213	196	182	159	141
25mm	514	463	422	388	358	333	292	260
32mm	900	818	750	692	643	600	529	474

For 38 mm conduit use the 32 mm factor x 1.4. For 50 mm conduit, use the 32 mm factor x 2.6. For 63 mm conduit use the 32 mm factor x 4.2.

Table 4.17 Trunking factors (terms)

Dimensions of trunking (mm x mm)			Factor
37.5	x	50	767
50	x	50	1037
25	x	75	738
37.5	x	75	1146
50	x	75	1555
75	x	75	2371
25	x	100	993
37.5	x	100	1542
50	x	100	2091
75	x	100	3189
100	x	100	4252

Example 4.9

The following single-core p.v.c. insulated cables are to be run in a conduit 6 m long with a double set: 8 x 1 mm², 4 x 2.5 mm² and 2 x 6 mm². Choose a suitable conduit size.

Consulting {Table 4.14} gives the following cable factors:

16 for 1 mm², 30 for 2.5 mm² and 58 for 6 mm²
Total cable factor is then
(8 × 16) + (4 × 30) + (2 × 58) = 128 + 120 + 116 = 364

The term 'bend' means a right angle bend or a double set.

{Table 4.16} gives a conduit factor for 20 mm conduit 6 m long with a double set as 233, which is less than 364 and thus too small. The next size has a conduit factor of 422 which will be acceptable since it is larger than 364. The correct conduit size is 25 mm diameter.

Example 4.10

The first conduit from a distribution board will be straight and 10 m long. It is to enclose 4 × 10 mm² and 8 × 4 mm² cables. Calculate a suitable size.

From {Table 4.14}, cable factors are 105 and 43 respectively. Total cable factor:

= (4 × 105) + (8 × 43) = 420 + 344 = 764

From {Table 4.16}, a 10 m long straight 25 mm conduit has a factor of 442. This is too small, so the next size, with a factor of 783 must be used.

The correct conduit size is 32 mm diameter.

Example 4.11

A 1.5 m straight length of conduit from a consumer's unit encloses ten 1.5 mm² and four 2.5 mm² solid conductor p.v.c. insulated cables. Calculate a suitable conduit size.

From {Table 4.15} (which is for short straight runs of conduit) total cable factor will be:

= (10 × 27) + (4 × 39) = 426

Table 4.16 shows that 20 mm diameter conduit with a factor of 460 will be necessary.

Example 4.12

A length of trunking is to carry eighteen 10 mm², sixteen 6 mm², twelve 4 mm², and ten 2.5 mm² stranded single thermosetting insulated cables. Calculate a suitable trunking size.

The total cable factor for trunking is calculated with data from {Table 4.14}.

18 × 10 mm² at 36.3	= 18 × 36.3	=	653.4
16 × 6 mm² at 22.9	= 16 × 22.9	=	366.4
12 × 4 mm² at 15.2	= 12 × 18.1	=	217.2
10 × 2.5 mm² at 11.4	= 10 × 13.9	=	139.0
	Total cable factor	=	1376.0

From the trunking factor {Table 4.17}, two standard trunking sizes have factors slightly greater than the cable factor, and either could be used . They are 50 mm × 75 mm at 1555, and 37.5 mm × 100 mm at 1542.

4.6 CONDUCTOR IDENTIFICATION [514, Appendix 7]

The colour identification for fixed wiring cables was changed to match those for Europe (HD 384.5.514) by the April 2004 amendments to BS 7671. Conductors may be coloured as stated, but other suitable methods of identification, including tapes, sleeves, discs or numbering, are acceptable. Methods may be mixed within an installation, but all new installations must use the new colour system. The original and the newer colours are shown in Table 4.18, together with the conductor markings for extensions and alterations.

Table 4.18 Summary of old and new conductor identification colours

Cable	Old colour	New colour	Colour marking for extensions and alterations
Flexible cables and cords			
Phase	Brown	Brown	Not applicable
Neutral	Blue	Blue	Not applicable
Protective	Green/yellow	Green/yellow	Not applicable
Fixed wiring, single-phase			
Phase	Red	Brown	Brown
Neutral	Black	Blue	Blue
Protective	Green/yellow	Green/yellow	Green/yellow
Fixed wiring, three-phase			
Phase 1	Red or L1	Brown or L1	Brown or L1
Phase 2	Yellow or L2	Black or L2	Black or L2
Phase 3	Blue or L3	Grey or L3	Grey or L3
Neutral	Black or 0	Blue or 0	Blue or 0
Protective	Green/yellow	Green/yellow	Green/yellow

Table 4.18 is extracted from Table 51 of BS 7671, which also includes data for d.c. circuits.

The question obviously arises of what to do when extending or altering an existing installation. With single-phase installations, there is no difficulty, because there is no ambiguity. A black neutral has simply been replaced by a blue, as has a red phase conductor by a brown. Where alterations or additions to existing installations are made, the old conductors at the interface should be marked with printed tape as shown by Table 4.19.

Table 4.19 Colour markings at the junctions of old and new three-phase wiring

Function	Old conductor insulation colour	Marking	New conductor Marking	Insulation colour
Phase of a.c.	Red	L1	Brown or L1	Brown
	Yellow	L2	Black or L2	Black
	Blue	L3	Grey or L3	Grey
Neutral	Black	N	Blue	Blue

If wiring alterations are made to an installation so that some of the wiring uses the new colours whilst there is also wiring to the old colours, a warning notice must be fixed at or near the appropriate distribution board(s) with the following wording:

CAUTION

This installation has wiring colours to two versions of BS 7671.
Great care should be taken before undertaking extension, alteration or repair that all conductors are correctly identified.

Where two-core brown and blue cables are used as phase switch wires (for example in two-way switching), the blue conductor must be marked brown or L at its terminations. It is likely that three core cables with brown, black and grey conductors will be used for intermediate switching. In this event, the black and grey conductors must both be marked brown or L.

Identification by coloured marking is not required for

i)   Concentric cables
ii) Metal sheath or armour used as a protective conductor
iii) Bare conductors where permanent marking is impractical
iv) Extraneous conductive parts where used as a protective conductor
v)   Exposed conductive parts where used as a protective conductor.

The question of danger arising from the use of the new colours will be asked. For single-phase installations, this is unlikely; we have all become very used to brown and blue conductors in flexible cords and cables. The use of a black conductor for the second line conductor of three-phase systems could be argued to present safety problems, because it has become familiar as a safe neutral conductor at earth potential. However, bearing in mind that black phase conductors must be identified at the interface with lettered markers (probably tape) and that the unskilled handyman is not likely to encounter three-phase wiring, the probability of an accident is (hopefully) remote. Another possible danger is that a blue phase conductor (old colours) may be mistaken for a neutral (new colours). Again, if the required markings are followed this is unlikely. Use of the Caution Notice above should also help to prevent confusion between old and newly coloured wiring systems.

The 'electrical' colour to distinguish electrical conduits from pipes of other services remains orange. Over-sheaths of mineral insulated cables are the same colour, which is also used to identify switchgear enclosures and trunking.

CHAPTER FIVE

Earthing

5.1 THE EARTHING PRINCIPLE

5.1.1 What is earthing? [54, Guidance Note 8]

The whole of the world may be considered as a vast conductor which is at reference (zero) potential. In the UK we refer to this as 'earth' whilst in the USA it is called 'ground'. People are usually more or less in contact with earth, so if other parts which are open to touch become charged at a different voltage from earth a shock hazard exists (see {3.4}). The process of earthing is to connect all these parts which could become charged to the general mass of earth, to provide a path for fault currents and to hold the parts as near as possible to earth potential. In simple theory this will prevent a potential difference between earth and earthed parts, as well as permitting the flow of fault current which will cause the operation of the protective systems.

The standard method of tying the electrical supply system to earth is to make a direct connection between the two. This is usually carried out at the supply transformer, where the neutral conductor (often the star point of a three-phase supply) is connected to earth using an earth electrode or the metal sheath and armouring of a buried cable. {Figure 5.1} shows such a connection. Lightning conductor systems must be bonded to the installation earth with a conductor no smaller in cross-sectional area than that of the earthing conductor.

Guidance Note 8, 'Earthing and Bonding' was published in the middle of 2007 and provides a huge store of data on the subject.

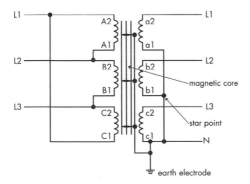

Fig 5.1 Three-phase delta/star transformer showing earthing arrangements

5.1.2 The advantages of earthing

The practice of earthing is widespread, but not all countries in the world use it. There is certainly a high cost involved, so there must be some advantages. In fact there are two. They are:

1 The whole electrical system is tied to the potential of the general mass of earth and cannot 'float' at another potential. For example, we can be fairly certain that the neutral of our supply is at, or near, zero volts (earth potential) and that the phase conductors of our standard supply differ from earth by 230 volts.

2 By connecting earth to metalwork not intended to carry current (an extraneous conductive part or an exposed conductive part) by using a protective conductor, a path is provided for fault current which can be detected and, if necessary, broken. The path for this fault current is shown in {Fig 5.2}.

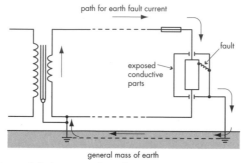

Fig 5.2 Path for earth fault current (shown by arrows)

5.1.3 The disadvantages of earthing

The two important disadvantages are:

1 Cost: the provision of a complete system of protective conductors, earth electrodes, etc. is very expensive.

2 Possible safety hazard: It has been argued that complete isolation from earth will prevent shock due to fault contact (previously called indirect contact) because there is no path for the shock current to return to the circuit if the supply earth connection is not made (see {Fig 5.3a}). This approach, however, ignores the presence of earth leakage resistance (due to imperfect insulation) and phase-to-earth capacitance (the insulation behaves as a dielectric). In many situations the combined impedance due to insulation resistance and earth capacitive reactance is low enough to allow a significant shock current (see {Fig 5.3b}).

a)   b)

Fig 5.3 Danger in an unearthed system
a) apparent safety: no obvious path for shock current
b) actual danger: shock current via stray resistance and capacitance

5.2 EARTHING SYSTEMS

5.2.1 System classification [411, 542.1]

The electrical installation does not exist on its own; the supply is part of the overall system. Although Electricity Supply Companies will often provide an earth terminal, they are under no legal obligation to do so. As far as earthing types are concerned, letter classifications are used.

The first letter indicates the type of supply earthing.

T indicates that one or more points of the supply are directly earthed (for example, the earthed neutral at the transformer).

I indicates either that the supply system is not earthed at all, or that the earthing includes a deliberately-inserted impedance, the purpose of which is to limit fault current. This method is not used for public supplies in the UK.

The second letter indicates the earthing arrangement in the installation.

T all exposed conductive metalwork is connected directly to earth.

N all exposed conductive metalwork is connected directly to an earthed supply conductor provided by the Electricity Supply Company.

The third and fourth letters indicate the arrangement of the earthed supply conductor system.

S neutral and earth conductor systems are quite separate.

C neutral and earth are combined into a single conductor.

A number of possible combinations of earthing systems in common use is indicated in the following subsections.

Protective conductor systems against lightning need to be connected to the installation earthing system to prevent dangerous potential differences. Where a functional earthing system is in use, the protective requirements of the earthing will take precedence over the functional requirements.

5.2.2 TT systems [411.5, 542.1]

This arrangement covers installations not provided with an earth terminal by the Electricity Supply Company. Thus it is the method employed by most (usually rural) installations fed by an overhead supply. Neutral and earth (protective) conductors must be kept quite separate throughout the installation, with the final earth terminal connected to an earth electrode (see {5.5}) by means of an earthing conductor. Effective earth connection is sometimes difficult. Because of this, socket outlet circuits must be protected by a residual current device (RCD) with an operating current not exceeding 30 mA {5.9}. {Fig 5.4} shows the arrangement of a TT earthing system.

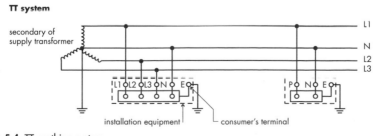

Fig 5.4 TT earthing system

5.2.3 TN-S system [411.4, 542.1]

This is probably the most usual earthing system in the UK, with the Electricity Supply Company providing an earth terminal at the incoming mains position. This earth terminal is connected by the supply protective conductor (PE) back to the star point (neutral) of the secondary winding of the supply transformer, which is also connected at that point to an earth electrode. The earth conductor usually takes the form of the armour and sheath (if applicable) of the underground supply cable. The system is shown diagrammatically in {Fig 5.5}.

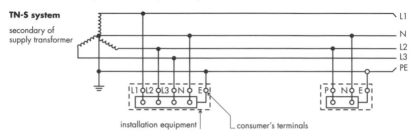

Fig 5.5 TN-S earthing system

5.2.4 TN-C-S system [411.4, 542.1]

In this system, the installation is TN-S, with separate neutral and protective conductors. The supply, however, uses a common conductor for both the neutral and the earth. This combined earth and neutral system is sometimes called the 'protective and neutral conductor' (PEN) or the 'combined neutral and earth' conductor (CNE). The system, which is shown diagrammatically in {Fig 5.6}, is most usually called the protective multiple earth (PME) system, which will be considered in greater detail in {5.6}.

5.2.5 TN-C system [537, 543]

This installation is unusual, because combined neutral and earth wiring is used in both the supply and within the installation itself. Where used, the installation will usually be the earthed concentric system, which can only be installed under the special conditions listed in {5.7}.

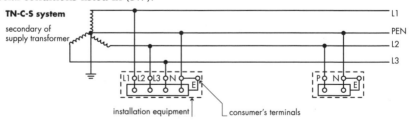

Fig 5.6 TN-C-S earthing system - protective multiple earthing

5.2.6 IT system [531.1.3, 538.1]

The installation arrangements in the IT system are the same for those of the TT system {5.2.2}. However, the supply earthing is totally different. The IT system can have an unearthed supply, or one which is not solidly earthed but is connected to earth through a current limiting impedance.

The total lack of earth in some cases, or the introduction of current limiting into the earth path, means that the usual methods of protection will not be effective. For this reason, IT systems are not allowed in the public supply system in the UK. An exception is in medical situations such as hospitals. Here it is recommended that an IT system is used for circuits supplying medical equipment that is intended to be used for life-support of patients. (see {8.19}). The method is also sometimes used where a supply for special purposes is taken from a private generator (see {8.21}).

5.3 EARTH FAULT LOOP IMPEDANCE

5.3.1 Principle [411]

The path followed by fault current as the result of a low impedance occurring between the phase conductor and earthed metal is called the earth fault loop. Current is driven through the loop impedance by the supply voltage. The extent of the earth fault loop for a TT system is shown in {Fig 5.7}, and is made up of the following labelled parts.

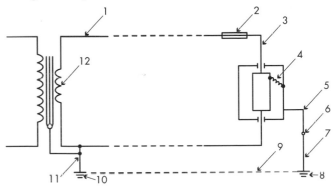

1 the phase conductor from the transformer to the installation
2 the protective device(s) in the installation
3 the installation phase conductors from the intake position to the fault
4 the fault itself (usually assumed to have zero impedance)
5 the protective conductor system
6 the main earthing terminal
7 the earthing conductor
8 the installation earth electrode
9 the general mass of earth
10 the Supply Company's earth electrode
11 the Supply Company's earthing conductor
12 the secondary winding of the supply transformer

Fig 5.7 The earth fault loop

For a TN-S system (where the Electricity Supply Company provides an earth terminal), items 8 to 10 are replaced by the PE conductor, which usually takes the form of the armouring (and sheath if there is one) of the underground supply cable.

For a TN-C-S system (protective multiple earthing) items 8 to 11 are replaced by the combined neutral and earth conductor.

For a TN-C system (earthed concentric wiring), items 5 to 11 are replaced by the combined neutral and earth wiring of both the installation and of the supply.

It is readily apparent that the impedance of the loop will probably be a good deal higher for the TT system, where the loop includes the resistance of two earth electrodes as well as an earth path, than for the other methods where the complete loop consists of metallic conductors.

5.3.2 The importance of loop impedance

The earth fault loop impedance can be used with the supply voltage to calculate the earth-fault current.

$$I_F = \frac{U_o}{Z_s}$$

where I_F = fault current, A

U_o = phase voltage, V

Z_s = loop impedance, Ω

For example, if a 230 V circuit is protected by a 15 A semi-enclosed fuse and has an earth-fault loop impedance of 1.6 Ω the earth-fault current in the event of a zero impedance earth fault will be:

$$I_F = \frac{U_o}{Z_s} = \frac{230 \text{ A}}{1.6} = 144 \text{ A}$$

This level of earth-fault current will cause the fuse to operate quickly. From {Fig 3.13} the time taken for the fuse to operate will be about 0.15 s. Any load current in the circuit will be additional to the fault current and will cause the fuse to operate slightly more quickly. However, such load current must not be taken into account when deciding disconnection time, because it is possible that the load may not be connected when the fault occurs.

Note that there is no such thing as a three-phase line/earth fault, although it is possible (but not likely) for three faults to occur on the three lines to earth simultaneously. As far as calculations for fault current are concerned, the voltage to earth for standard UK supplies is always 230 V, for both single-phase and three-phase systems. Thus the Tables of maximum earth-fault loop impedance which will be given in {5.3.4} apply both to single- and to three-phase systems.

5.3.3 The resistance/impedance relationship

Resistance, measured in ohms (Ω), is the property of a conductor to limit the flow of current through it when a voltage is applied.

$$I = \frac{U}{R}$$

where I = current, A

U = applied voltage, V

R = circuit resistance, Ω

Thus, a voltage of one volt applied to a resistance of one ohm results in a current of one ampere.

When the supply voltage is alternating, a second effect, known as reactance (symbol X) has to be considered. It applies only when the circuit includes inductance and/or capacitance, and its value, measured in ohms, depends on the frequency of the supply as well as on the values of the inductance and/or the capacitance concerned. For almost all installation work in the UK the frequency is constant at 50 Hz. Thus, inductive reactance is directly proportional to inductance and capacitive reactance is inversely proportional to capacitance.

$$X_L = 2\pi f L \quad \text{and} \quad X_C = \frac{1}{2\pi f_C}$$

where X_L = inductive reactance (Ω)
X_C = capacitive reactance (Ω)
π = the mathematical constant (3.142)
f = the supply frequency (Hz)
L = circuit inductance (H)
C = circuit capacitance (F)

Resistance (R) and reactance (X_L or X_C) in series add together to produce the circuit impedance (symbol Z), but not in a simple arithmetic manner. Impedance is the effect that limits alternating current in a circuit containing reactance as well as resistance.

$$Z = \frac{U}{I}$$

where Z = impedance (Ω)
U = applied voltage (V)
I = current (A)

It follows that a one volt supply connected across a one ohm impedance results in a current of one ampere.

When resistance and reactance are added this is done as if they were at right angles, because the current in a purely reactive circuit is 90° out of phase with that in a purely resistive circuit. The relationships between resistance, reactance and impedance are:

$$Z = \sqrt{(R^2 + X_L^2)} \quad \text{and} \quad Z = \sqrt{(R^2 + X_C^2)}$$

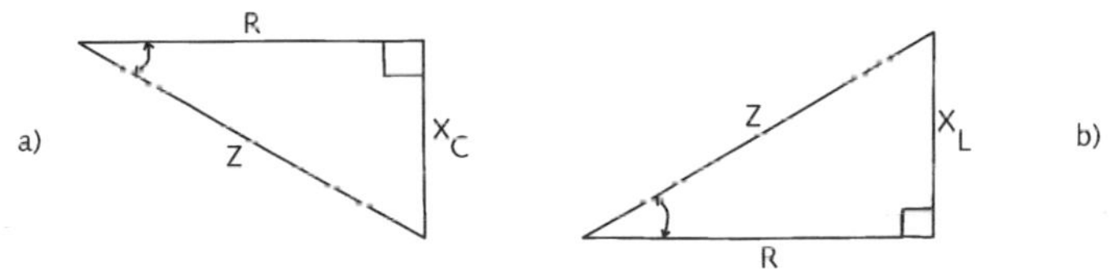

Fig 5.8 Impedance diagrams
a) resistive and capacitive circuit   b) resistive and inductive circuit

These relationships can be shown in the form of a diagram applying Pythagoras' theorem as shown in {Fig 5.8}. The two diagrams are needed because current lags voltage in the inductive circuit, but leads it in the capacitive. The angle between the resistance R and the impedance Z is called the circuit phase angle, given the symbol ø (Greek 'phi'). If voltage and current are both sinusoidal, the cosine of this angle, cos ø, is the circuit power factor, which is said to be lagging for the inductive circuit, and leading for the capacitive.

In practice, all circuits have some inductance and some capacitance associated with them. However, the inductance of cables only becomes significant when they have a cross-sectional area of 25 mm² and greater. Remember that the higher the earth fault loop impedance the smaller the fault current will be. Thus, if simple arithmetic is used to add resistance and reactance, and the resulting impedance is low enough to open the protective device quickly enough, the circuit will be safe. This is because the Pythagorean addition will always give lower values of impedance than simple addition.

For example, if resistance is 2 Ω and reactance 1 Ω simple arithmetic addition gives

$$Z = R + X = 2 + 1 = 3\,\Omega$$

and correct addition gives

$$Z = \sqrt{(R^2 + X^2)}$$
$$= \sqrt{(2^2 + 1^2)} = \sqrt{5} = 2.24\,\Omega$$

If 3 Ω is acceptable, 2.24 Ω will allow a larger fault current to flow, which will operate the protective device more quickly and is thus even more acceptable.

5.3.4 Earth-fault loop impedance values [411.3 to 411.5]

The over-riding requirement is that sufficient fault current must flow in the event of an earth fault to ensure that the protective device cuts off the supply before dangerous shock can occur. For normal 230 V systems, there are two levels of maximum disconnection time. These are:

For 230 V socket outlet circuits where equipment could be tightly grasped: 0.4 s, or for 400 V equipment 0.2 s. Note that the 17th Edition now requires **most** socket outlets with ratings up to 32 A to be protected by a 30 mA rated RCD so that disconnection time should not exceed 150 ms. For TT systems, this time is reduced to 0.2 s for 230 V supplies or to 0.07 s for 400 V systems, although 400 V supplies are not likely in a TT system.

The maximum disconnection time of 5 s also applies to feeders and sub-mains. It must be appreciated that the longest disconnection times for protective devices, leading to the longest shock times and the greatest danger, will be associated with the lowest levels of fault current, and not, as is commonly believed, the highest levels.

Where the voltage is other than 230 V, [Table 41.1] gives a range of disconnection times for socket outlet circuits, of which the lowest is 0.1 s for voltages exceeding 400 V. In general, the requirement is that if a fault of negligible impedance occurs between a phase and earth, the earth-fault loop impedance must not be greater than the value calculated from:

$$Z_s < \frac{U_o}{I_a}$$

where Z_s = the earth fault loop impedance (Ω)

U_o = the system voltage to earth (V)

I_a = the current causing automatic disconnection (operation of the protective device) in the required time (A).

The earth fault loop impedance values shown in [Tables 41.2, 41.3 and 41.4] depend on the supply voltage and assume, as shown in the Tables, a value of 230 V. Whilst it would appear that 230 V is likely to be the value of the supply voltage in Great Britain for the foreseeable future, it is not impossible that different values may apply. In such a case, the tabulated value for earth fault loop impedance should be modified using the formula:

$$Z_s = Z_t \times \frac{U}{U_{230}}$$

where Z_s is the earth fault loop impedance required for safety (Ω)

Z_t is the tabulated value of earth fault loop impedance (Ω)

U   is the actual supply voltage (V)

U_{230} is the supply voltage assumed in the Table (V).

As an alternative to this calculation, a whole series of maximum values of earth fault loop impedance is given in {Table 5.1} (from [Table 41B]) for disconnection within 0.4 s. The reader should not think that these values are produced in some mysterious way – all are easily verified using the characteristic curves {Figs 3.13 to 3.18}.

For example, consider a 20 A HBC fuse to BS88 used in a 230 V system. The fuse characteristic is shown in {Fig 3.15}, and indicates that disconnection in 0.4 s requires a current of about 130 A. It is difficult (if not impossible) to be precise about this value of current, because it is between the 100 A and 150 A current graduations. Using these values,

$$Z_s = \frac{U_o}{I_a} = \frac{230}{130} = 1.77 \ \Omega$$

Reference to {Table 5.1} shows that the stated value is 1.77 Ω. {Tables 5.1 and 5.2} give maximum earth-fault loop impedance values for fuses and for miniature circuit breakers to give a minimum disconnection time of 0.4 s in the event of a zero impedance fault from phase to earth.

Table 5.1 Maximum earth fault loop impedance for 230 V socket outlet circuits protected by fuses – 0.4 s disconnection time
(from [Table 41.2] of BS 7671: 2008)

Fuse rating (A)	Maximum earth-fault loop impedance (Ω) Cartridge BS 88	Cartridge BS 1361	Cartridge BS 1362	Semi-enclosed BS 3036
5	—	10.45	—	9.58
6	8.52	—	—	—
10	5.11	—	—	—
13	—	—	2.42	—
15	—	3.28	—	2.55
16	2.70	—	—	—
20	1.77	1.70	—	1.77
25	1.44	—	—	—
30	—	1.15	—	1.09
32	1.04	—	—	—

Note that the figures in {Tables 5.1 to 5.3} are lower than those in the corresponding Tables in the 16th Edition since they are based on a supply voltage of 230 V rather than 240 V.

A new [Table 41.5] is included for the first time in the 17th Edition giving maximum earth fault loop impedance values for correct operation of non-delayed RCDs. As would be expected, values are very high compared with those in {Tables 5.1 to 5.3}. See {Table 5.4}.

Table 5.2 Maximum earth-fault loop impedance for 230 V socket outlet circuits protected by miniature circuit breakers – 0.4 s and 5 s disconnection times (from [Table 41.3] of BS 7671: 2008)

MCB rating (A)	Maximum earth-fault loop impedance (Ω) Type B	Type C	Type D
6	7.67	3.83	1.92
10	4.60	2.30	1.15
16	2.87	1.44	0.72
20	2.30	1.15	0.57
25	1.84	0.92	0.46
32	1.44	0.72	0.36
40	1.15	0.57	0.29
50	0.92	0.46	0.23

Table 5.3 Maximum earth-fault loop impedance for 230 V distribution circuits protected by fuses – 5.0 s disconnection time (from [Table 41.4] of BS 7671: 2008)

Fuse rating (A)	Maximum earth-fault loop impedance (Ω) Cartridge BS 88	Cartridge BS 1361	Cartridge BS 1362	Semi-enclosed BS 3036
5	—	16.4	—	17.7
6	13.5	—	—	—
10	7.42	—	—	—
13	—	—	3.83	—
15	—	5.00	—	5.35
16	4.18	—	—	—
20	2.91	2.80	—	3.83
25	2.30	—	—	—
30	—	1.84	—	2.64
32	1.84	—	—	—

Table 5.4 Maximum earth-fault loop impedance to ensure correct operation of non-delayed residual current devices for final circuits not exceeding 32 A (from [Table 41.5] of BS 7671: 2008)

Rated residual operating current (mA)	Maximum earth-fault loop impedance (Ω) From 50 V to 230 V	From 230 V to 400 V
30	1667	1533
100	500	460
300	167	153
500	100	92

The resistance of the earth electrode should be as low as possible and never more than 100 Ω.

The severity of the electric shock received when there is a phase to earth fault (fault protection) depends entirely on the impedance of the circuit protective conductor. We saw in {3.4.3} and {Fig 3.8} how the volt drop across the protective conductor is applied to the person receiving the shock. Since this volt drop is equal to fault current times protective conductor impedance, if the protective conductor has a lower impedance the shock voltage will be less. Thus it can be sustained for a longer period without extreme danger.

Socket outlet circuits can therefore have a disconnection time of up to 5 s provided that the circuit protective conductor impedances are no higher than shown

in {Table 5.3} for various types of protection; note however, that the new requirement of the 17th Edition that most socket outlets should be protected by an RCD
means that disconnection times greater than 150 ms are unlikely.

The reasoning behind this set of requirements becomes clearer if we take an
example. {Table 5.3} shows that a 40 A cartridge fuse to BS 88 must have an associated protective conductor impedance of no more than 0.26 Ω if it is to comply.
Now look at the time/current characteristic for the fuse {Fig 3.15} from which we
can see that the current for operation in 5 s is about 170 A. The maximum volt drop
across the conductor (the shock voltage) is thus 170 x 0.26 or 44.2 V. Application
of the same reasoning to all the figures gives shock voltages of less than 50 V. This
limitation on the impedance of the CPC is of particular importance in TT systems
where it is likely that the resistance of the earth electrode to the general mass of
earth will be high.

The breaking time of 5 s also applies to fixed equipment, so the earth-fault loop
impedance values can be higher for these circuits, as well as for distribution circuits.
For fuses, the maximum values of earth-fault loop impedance for fixed equipment
are given in {Table 5.4}.

No separate values are given for miniature circuit breakers. Examination of the
time/current characteristics {Figs 3.16 to 3.18} will reveal that there is no change at
all in the current causing operation between 0.4 s and 5 s. In this case, the values
given in {Table 5.2} can be used for fixed equipment as well as for socket outlet
circuits. An alternative is to calculate the loop impedance as described above.

5.3.5 Protective conductor impedance [54]

It has been shown in the previous sub-section how a low-impedance protective
conductor will provide safety from shock in the event of a fault to earth. This
method can only be used where it is certain that the shock victim can never be in
contact with conducting material at a different potential from that of the earthed
system in the zone he occupies. Thus, all associated exposed or extraneous parts
must be within the equipotential zone (see {5.4}) When overcurrent protective
devices are used as protection from electric shock, the protective conductor must be
in the same wiring system as, or in close proximity to, the live conductors. This is
intended to ensure that the protective conductor is unlikely to be damaged in an
accident without the live conductors also being cut.

{Figure 5.9} shows a method of measuring the resistance of the protective
conductor, using a line conductor as a return and taking into account the different
cross-sectional areas of the phase and the protective conductors.

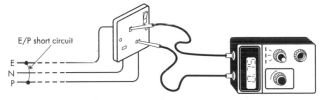

Fig 5.9 Measurement of protective conductor resistance

Taking the cross-sectional area of the protective conductor as A_p and that of the line
(phase or neutral) conductor as A_L, then

$$Rp = \text{resistance reading} \times \frac{A_L}{A_L + A_p}$$

For example, consider a reading of 0.72 Ω obtained when measuring a circuit in the way described and having 2.5 mm^2 line conductors and a 1.5 mm^2 protective conductor. The resistance of the protective conductor is calculated from:

$$R_p = R \times \frac{A_L}{A_L + A_p} = \frac{0.72 \times 2.5}{2.5 + 1.5} = \frac{0.72 \times 2.5}{4.0} = 0.45\,\Omega$$

5.3.6 Maximum circuit conductor length

The complete earth-fault loop path is made up of a large number of parts as shown in {Fig 5.7}, many of which are external to the installation and outside the control of the installer. These external parts make up the external loop impedance (Z_E). The rest of the earth-fault loop impedance of the installation consists of the impedance of the phase and protective conductors from the intake position to the point at which the loop impedance is required.

Since an earth fault may occur at the point farthest from the intake position, where the impedance of the circuit conductors will be at their highest value, this is the point which must be considered when measuring or calculating the earth-fault loop impedance for the installation. Measurement of the impedance will be considered in {7.6.2}. Provided that the external fault loop impedance value for the installation is known, total impedance can be calculated by adding the external impedance to that of the installation conductors to the point concerned. The combined resistance of the phase and protective conductors is known as $R_1 + R_2$. The same term is sometimes used for the combined resistance of neutral and protective conductors (see {7.4.1}). In the vast majority of cases phase and neutral conductors have the same cross-sectional area and hence the same resistance. For the majority of installations, these conductors will be too small for their reactance to have any effect (below 25 mm^2 cross-sectional area reactance is very small), so only their resistances will be of importance. This can be measured by the method indicated in {Fig 5.9}, remembering that this time we are interested in the combined resistance of phase and protective conductors, or can be calculated if we measure the cable length and can find data concerning the resistance of various standard cables. These data are given here as {Table 5.5}.

Table 5.5 Resistance per metre of copper conductors at 20°C for calculation of $R_1 + R_2$

Conductor cross-sectional area (mm²)	Resistance per metre run (mΩ/m)
1.0	18.10
1.5	12.10
2.5	7.41
4.0	4.61
6.0	3.08
10.0	1.83
16.0	1.15
25.0	0.727

Note that to allow for the increase in resistance with increased temperature under fault conditions the values of Table 5.5 must be multiplied by 1.2 for pvc insulated cables (see {Table 7.7}

The resistance values given in {Table 5.5} are for conductors at 20°C. Under fault conditions the high fault current will cause the temperature of the conductors to rise and result in an increase in resistance. To allow for this changed resistance, the values in {Table 5.5} must be multiplied by the appropriate correction factor from Table {7.7}. It should be mentioned that the practice which has been adopted here of adding impedance and resistance values arithmetically is not strictly correct. Phasor addition is the only perfectly correct method since the phase angle associated with resistance is likely to be different from those associated with impedance, and in addition impedance phase angles will differ from one another. However, if the phase angles are similar, and this will be so in the vast majority of cases where electrical installations are concerned, the error will be acceptably small. It is often assumed that higher conductor temperatures are associated with the higher levels of fault current. In most cases this is untrue. A lower fault level will result in a longer period of time before the protective device operates to clear it, and this often results in higher conductor temperature.

Example 5.1
A 7 kW shower heater is to be installed in a house fed with a 230 V TN-S supply system with an external loop impedance (Z_E) of 0.8 Ω. The heater is to be fed from a 32 A BS 88 cartridge fuse. Calculate a suitable size for the p.v.c. insulated and sheathed cable to be used and determine the maximum possible length of run for this cable. It may be assumed that the cable will not be subject to any correction factors and is clipped direct to a heat conducting surface throughout its run.

First calculate the circuit current.

$$I = \frac{P}{U} = \frac{7000}{230} \text{ A} = 30.4 \text{ A}$$

Next, select the cable size from {Table 4.7} (which is based on [Table 4D2A]) from which we can see that 4 mm² cable of this kind clipped direct has a rating of 36 A. The 2.5 mm² rating of 27 A is not large enough. We shall assume that a 2.5 mm² protective conductor is included within the sheath of the cable. Now we must find the maximum acceptable earth-fault loop impedance for the circuit. Since this is a shower, a maximum disconnection time of 0.4 s will apply, so we need to consult {Table 5.1}, which gives a maximum loop impedance of 1.09 Ω for this situation. Since the external loop impedance is 0.8 Ω we can calculate the maximum resistance of the cable.

R cable = max. loop impedance – external impedance
 = 1.09 – 0.8 Ω = 0.29 Ω

This assumes that resistance and impedance phase angles are identical, which is not strictly the case. However, the difference is unimportant.

The phase conductor (4 mm²) has a resistance of 4.61 mΩ/m and the 2.5 mm² protective conductor 7.41 mΩ/m, so the combined resistance per metre is 4.61 + 7.41 = 12.0 mΩ/m. The cable will get hot under fault conditions, so we must apply the multiplier of 1.2.

Effective resistance of cable per metre = 12.0 × 1.2 mΩ/m = 14.4 mΩ/m

The maximum length of run in metres is thus the number of times 14.4 mΩ will divide into 0.29 Ω.

Maximum length of run = $\dfrac{0.29 \times 1000}{14.4}$ m = 20.1 m

This is not quite the end of our calculation, because we must check that this length of run will not result in an excessive volt drop. From {Table 4.7} based on [Table 4D2B] a 4 mm^2 cable of this kind gives a volt drop of 11 mV/A/m.

Volt drop = 11 × circuit current (A) × length of run (m) divided by 1000

$$\text{Volt drop} = \frac{11 \times 30.4 \times 20.1}{1000} \text{ V} = 6.72 \text{ V}$$

Since the permissible volt drop is 5% of 230 V or 11.5 V, this length of run is acceptable.

5.4 PROTECTIVE CONDUCTORS

5.4.1 Earthing conductors [542, 543]

The earthing conductor is commonly called the earthing lead. It joins the installation earthing terminal to the earth electrode or to the earth terminal provided by the Elecricity Supply Company. It is a vital link in the protective system, so care must be taken to see that its integrity will be preserved at all times. Aluminium conductors and cables may now be used for earthing and bonding, but great care must be taken when doing so to ensure that there will be no problems with corrosion or with electrolytic action where they come into contact with other metals.

Where the final connection to the earth electrode or earthing terminal is made there must be a clear and permanent notice **Safety Electrical Connection – Do not remove** (see {Fig 5.17}). Where a buried earthing conductor is not protected against mechanical damage but is protected against corrosion by a sheath, its minimum size must be 16 mm^2, whether made of copper or coated steel. If it has no corrosion protection, minimum sizes for mechanically unprotected earthing conductors are 25 mm^2 for copper and 50 mm^2 for coated steel.

If not protected against corosion the latter sizes again apply, whether protected from mechanical damage or not.

Earthing conductors, as well as protective and bonding conductors, must themselves be protected against corrosion. Probably the most common type of corrosion is electrolytic, which is an electro-chemical effect between two different metals when a current passes between them whilst they are in contact with each other and with a weak acid. The acid is likely to be any moisture which has become contaminated with chemicals carried in the air or in the ground. The effect is small on ac supplies because any metal removed whilst current flows in one direction is replaced as it reverses in the next half cycle. For dc systems, however, it will be necessary to ensure that the system remains perfectly dry (a very difficult task) or to use the 'sacrificial anode' principle.

A main earth terminal or bar must be provided for each installation to collect and connect together all protective and bonding conductors. It must be possible to disconnect the earthing conductor from this terminal for test purposes, but only by the use of a tool. This requirement is intended to prevent unauthorised or unknowing removal of protection.

5.4.2 Protective conductor types [413]

The circuit protective conductor (increasingly called the 'c.p.c.') is a system of conductors joining together all exposed conductive parts and connecting them to the main earthing terminal. Strictly speaking, the term includes the earthing conductor as well as the equipotential bonding conductors.

The circuit protective conductor can take many forms, such as:

1 a separate conductor which must be green/yellow insulated if equal to or less than 10 mm² cross-sectional area. There are three cases where a separate protective conductor must be used
   a) to connect the earthing terminal of an accessory to its metal box or trunking
   b) through a metal or other flexible conduit
   c) in a final circuit feeding equipment with normal protective current exceeding 10 mA (see {8.8})
2 a conductor included in a sheathed cable with other conductors
3 the metal sheath and/or armouring of a cable
4 conducting cable enclosures such as conduit or trunking
5 exposed conductive parts, such as the conducting cases of equipment

This list is by no means exhaustive and there may be many other items forming parts of the circuit protective conductor as indicated in {Fig 5.10}. Note that gas or oil pipes must not be used for this purpose, because of the possible future change to plastic (non-conducting) pipes.

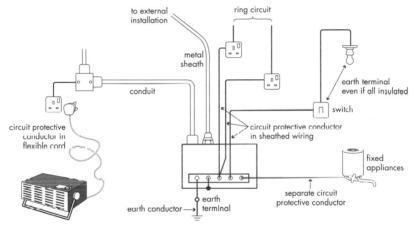

Fig 5.10 Some types of circuit protective conductor

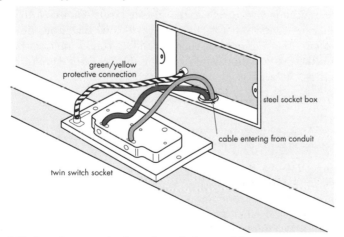

Fig 5.11 Protective connection for socket outlet in a conduit system

It is, of course, very important that the protective conductor remains effective throughout the life of the installation. Thus, great care is needed to ensure that steel conduit used for the purpose is tightly jointed and unlikely to corrode. The difficulty of ensuring this point is leading to the increasing use of a c.p.c. run inside the conduit with the phase conductors. Such a c.p.c. will, of course, always be necessary where plastic conduits are used. Where an accessory is connected to a system (for example, by means of a socket outlet) that uses conduit as its c.p.c., the appliance (or socket outlet) earthing terminal must be connected by a separate conductor to the earth terminal of the conduit box (see {Fig 5.11}) where there is doubt about the effectiveness of the earthing connections formed by screws between the accessory and the metal box. This connection will ensure that the accessory remains properly earthed even if the screws holding it into the box become loose, damaged or corroded. A separate protective conductor will be needed where flexible conduit is used, since this type of conduit cannot be relied upon to maintain a low resistance conducting path (see {Fig 5.12}).

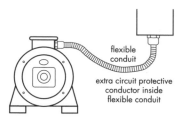

Fig 5.12 Separate additional protective conductor with flexible conduit

The continuity of metal trunking when used as a protective conductor is sometimes a cause for concern. In such a case, copper links can be used to join adjacent trunking lengths or accessories.

5.4.3 Bonding conductors [411, 544]
The purpose of the protective conductors is to provide a path for earth fault current so that the protective device will operate to remove dangerous potential differences, which are unavoidable under fault conditions, before a dangerous shock can be delivered. Equipotential bonding serves the purpose of ensuring that the earthed metalwork (exposed conductive parts) of the installation is connected to other metalwork (extraneous conductive parts) to ensure that no dangerous potential differences can occur. The resistance of such a bonding conductor must be low enough to ensure that its volt drop when carrying the operating current of the protective device never exceeds 50 V.

Thus $R < \dfrac{50}{I_a}$ NOTE that the symbol < means 'is smaller than'.

where   R is the resistance of the bonding conductor
I_a is the operating current of the protective device.

Two types of equipotential bonding conductor are specified.

1 Main equipotential bonding conductors
These conductors connect together the installation earthing system and the metalwork of other services such as gas and water. This bonding of service pipes must be effected as close as possible to their point of entry to the building, as shown in {Fig

5.13}. Metallic sheaths of telecommunication cables must be bonded, but the consent of the owner of the cable must be obtained before doing so. The minimum size of bonding conductors is related to the size of the main supply conductors (the tails) and is given in {Table 5.6}.

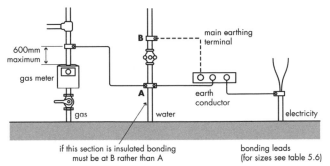

Fig 5.13 Main bonding connections

2 Supplementary bonding conductors

These conductors connect together extraneous conductive parts - that is, metalwork which is not associated with the electrical installation but which may provide a conducting path giving rise to shock. The object is to ensure that potential differences in excess of 50 V between accessible metalwork cannot occur; this means that the resistance of the bonding conductors must be low (see {Table 5.6}). {Figure 5.14} shows some of the extraneous metalwork in a bathroom which must be bonded.

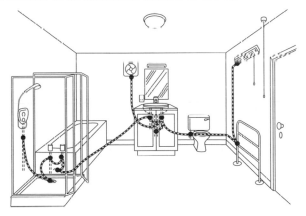

Fig 5.14 Supplementary bonding in a bathroom

Table 5.6 Supplementary bonding conductor sizes

Circuit protective conductor size (mm²)	Supplementary bonding conductor size (mm²) not protected	mechanically protected
1.0	4.0	2.5
1.5	4.0	2.5
2.5	4.0	2.5
4.0	4.0	2.5
6.0	4.0	2.5
10.0	4.0	2.5

The cross-sectional areas required for supplementary bonding conductors are shown in {Table 5.6}. Where connections are between extraneous parts only, the conductors may be 2.5 mm² if mechanically protected or 4 mm² if not protected. If the circuit protective conductor is larger than 10 mm², the supplementary bonding conductor must have at least half this cross-sectional area. Supplementary bonding conductors of less than 16 mm² cross sectional area must not be aluminium. {Fig 5.15} shows the application of a supplementary bonding conductor to prevent the severe shock which could otherwise occur between the live case of a faulty electric kettle and an adjacent water tap.

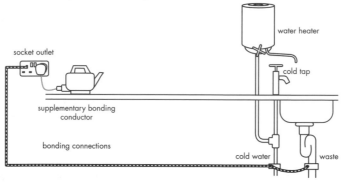

Fig 5.15 Supplementary bonding conductor in a kitchen

There will sometimes be doubt if a particular piece of metalwork should be bonded. The answer must always be that bonding will be necessary if there is a danger of severe shock when contact is made between a live system and the metal work in question. Thus if the resistance between the metalwork and the general mass of earth is low enough to permit the passage of a dangerous shock current, then the metalwork must be bonded.

The question can be resolved by measuring the resistance (R_x) from the metalwork concerned to the main earthing terminal. Using this value in the formula:

$$I_b = \frac{U_o}{R_p + R_x}$$

will allow calculation of the maximum current likely to pass through the human body where:

I_b is the shock current through the body (A)

U_o is the voltage of the supply (V)

R_p is the resistance of the human body (Ω) and

R_x is the measured resistance from the metalwork concerned to the main earthing terminal (Ω).

The resistance of the human body, R_p can in most cases be taken as 1000 Ω although 200 Ω would be a safer value if the metalwork in question can be touched by a person in a bath. Although no hard and fast rules are possible for the value of a safe shock current, I_b, it is probable that 10 mA is seldom likely to prove fatal. Using this value with 230 V for the supply voltage, U_o, and 1000 Ω as the human body resistance, R_p, the minimum safe value of R_x calculates to 22 kΩ. If the safer values of 5 mA for I_b and 200 Ω for R_p are used, the value of R_x would be 45.8 kΩ for a 230 V supply.

To sum up, when in doubt about the need to bond metalwork, measure its resistance to the main earthing terminal. If this value is 50 kΩ or greater, no bonding is necessary. In a situation where a person is not wet, bonding could be ignored where the resistance to the main earthing terminal is as low as 25 kΩ. To reduce the possibility of bonding conductors being disconnected by those who do not appreciate their importance, every installation should be provided with a label like that shown in {Fig. 5.17}.

Plastic pipes should not be bonded. It has been shown that the electrical resistance of tap water in such a pipe is too high to pose a shock danger; so metal taps fed by such pipes do not require bonding. Electrical equipment fed by plastic piping, such as an electrical shower unit, must be bonded.

5.4.4 Protective conductor cross-section assessment [543.1]

A fault current will flow when an earth fault occurs. If this current is large enough to operate the protective device quickly, there is little danger of the protective conductor and the exposed conductive parts it connects to earth being at a high potential to earth for long enough for a dangerous shock to occur. The factors determining the fault current are the supply voltage and the earth-fault loop impedance (see {5.3}).

The earth fault results in the protective conductors being connected in series across the supply voltage {Fig 5.16}. The voltage above earth of the earthed metalwork (the voltage of the junction between the protective and phase conductors) at this time may become dangerously high, even in an installation complying with the Regulations. The people using the installation will be protected by the ability of the fuse or circuit breaker in a properly designed installation to cut off the supply before dangerous shock damage results.

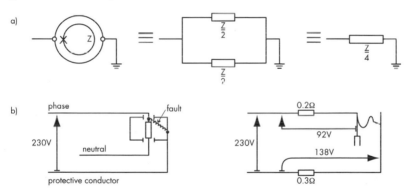

Fig 5.16 The effect of protective conductor resistance on shock voltage
a) effective resistance of a ring circuit protective conductor
b) potential differences across healthy protective conductor in the event of an earth fault

Remember that lower fault levels result in a longer time before operation of the protective device. Since the cross-sectional area of the protective conductor will usually be less than that of live conductors, its temperature, and hence its resistance, will become higher during the fault, so that the shock voltage will be a higher proportion of the supply potential (see {Fig 5.16}).

{Fig 5.16} shows the circuit arrangements, with some typical phase- and protective-conductor resistances. In this case, a shock voltage of 138 V will be applied to a person in contact with earthed metal and with the general mass of earth. Thus, the supply must be removed very quickly. The actual voltage of the shock depends directly on the relationship between the phase conductor resistance and the protective conductor resistance. If the two are equal, exactly half the supply voltage will appear as the shock voltage.

For socket outlet circuits, where the shock danger is highest, the maximum protective conductor resistance values of {Table 5.3} will ensure that the shock voltage never exceeds the safe value of 50 V. If the circuit concerned is in the form of a ring, one quarter of the resistance of the complete protective conductor round the ring must not be greater than the {Table 5.3} figure. The reason for this is shown in {Fig 5.16(a)}. This assumes that the fault will occur exactly at the mid point of the ring. If it happens at any other point, effective protective conductor resistance is lower, and safer, than one quarter of the total ring resistance.

{Table 5.7} allows selection (rather than calculation) of sizes for earthing and bonding conductors.

Note that Regional Electricity Companies may require a minimum size of earthing conductor of 16 mm² at the origin of the installation. Always consult them before designing an installation.

5.4.5 Protective conductor cross-section calculation [543.1]

Table 5.7 Main equipotential conductor sizes for TN-S and TN-C-S supplies
Derived from [Table 54.7] of BS 7671:2008

Neutral conductor c.s.a. (mm²)	Main equipotential bonding conductor for PME supplies c.s.a. (mm²)	Main equipotential bonding conductor c.s.a. (mm²)
4	10	10
6	10	10
10	10	10
16	10	10
25	10	10
35	10	10
50	16	16
70	25	25

The c.s.a. of the circuit protective conductor (cpc) is of great importance since the level of possible shock in the event of a fault depends on it (as seen in {5.4.4}). Safety could always be assured if we assessed the size using {Table 5.7} as a basis. However, this would result in a more expensive installation than necessary because we would often use protective conductors which are larger than those found to be acceptable by calculation. For example, twin with cpc insulated and sheathed cables larger than 1 mm² would be ruled out because in all other sizes the cpc is smaller than required by {Table 5.7}. In very many cases, calculation of the cpc size will show that a smaller size than that detailed in {5.4.4} is perfectly adequate. The formula to be used is:

$$S = \frac{\sqrt{(I_a^2 t)}}{k} \text{ mm}^2$$

where   S   is the minimum protective conductor cross-sectional area (mm²)
        I_a is the fault current (A)
        t   is the opening time of the protective device (s)
        k   is a factor depending on the conductor material and insulation, and the initial and maximum insulation temperatures.

This is the same formula as in {3.7.3}, the adiabatic equation, but with a change in the subject. To use it, we need to have three pieces of information, I_a, t and k.

1 To find I_a

Since $I_a = \dfrac{U_o}{Z_s}$ we need values for U_o and for Z_s.

U_o is simply the supply voltage, which in most cases will be 230V

Z_s is the earth-fault loop impedance assuming that the fault has zero impedance.

Since we must assume that we are at the design stage, we cannot measure the loop impedance and must calculate it by adding the loop impedance external to the installation (Z_E) to the resistance of the conductors to the furthest point in the circuit concerned. This technique was used in {5.3.6}.

Thus, $Z_s = Z_E + R_1 + R_2$ where R_1 and R_2 are the resistances of the phase and protective conductors respectively from {Table 5.5}.

2 To find t

We can find t from the time/current characteristics of {Figs 3.13 to 3.18} using the value of I_a already calculated above. For example, if the protective device is a 20 A miniature circuit breaker type B and the fault current is 1000 A, we shall need to consult {Fig 3.16}, when we can read off that operation will be in 0.1 s (100 ms). (It is of interest here to notice that if the fault current had been 80 A the opening time could have been anything from 0.1 s to 20 s, so the circuit would not have complied with the required opening times).

3 To find k

k is a constant, which we cannot calculate but must obtain from a suitable table of values. Some values of k for typical protective conductors are given in {Table 5.8}.

It is worth pointing out here that correctly installed steel conduit and trunking will always meet the requirements of the Regulations in terms of protective conductor impedance.

Although appearing a little complicated, calculation of acceptable protective conductor size is worth the trouble because it often allows smaller sizes than those shown in {Table 5.7}.

Table 5.8 Values of k for protective conductor in a cable or bunched with cables (From [Table 54.3] of BS 7671: 2008)

Conductor Material	Insulation material	Assumed initial temperature (°C)	Limiting final temperature (°C)	k
Copper	70°C thermoplastic (p.v.c.)	70	160	115
	90°C thermoplastic (p.v.c.)	90	160	100
	85°C thermosetting (rubber)	85	220	134
	90°C thermosetting	90	250	143
Aluminium	70°C thermoplastic (p.v.c.)	70	160	76
	90°C thermoplastic (p.v.c.)	90	160	66
	85°C thermosetting (rubber)	85	220	89
	90°C thermosetting	90	250	94

Further values for k applying to protective conductors will be found in [Tables 54.2 to 54.5] of BS 7671: 2008.

Example 5.2

A load takes 30 A from a 230 V single phase supply and is protected by a 32 A HBC fuse to BS 88. The wiring consists of 4 mm² single core p.v.c. insulated cables run in trunking, the length of run being 18 m. The earth-fault loop impedance external to the installation is assessed as 0.7 Ω. Calculate the cross-sectional area of a suitable p.v.c. insulated protective conductor.

This is one of those cases where we need to make an assumption of the answer to the problem before we can solve it. Assume that a 2.5 mm² protective conductor will be acceptable and calculate the combined resistance of the phase and protective conductors from the origin of the installation to the end of the circuit. From {Table 5.5}, 2.5 mm² cable has a resistance of 7.4 mΩ/m and 4 mm² a resistance of 4.6 mΩ/m. Both values must be multiplied by 1.2 to allow for increased resistance as temperature rises due to fault current.

$$\text{Thus, } R_1 + R_2 = \frac{(7.4 + 4.6) \times 1.2 \times 18}{1000} \ \Omega = \frac{12.0 \times 1.2 \times 18}{1000} \ \Omega = 0.26 \ \Omega$$

This conductor resistance must be added to external loop impedance to give the total earth fault loop impedance.

$$Z_s = Z_E + R_1 + R_2 = 0.7 + 0.26 \ \Omega = 0.96 \ \Omega$$

We can now calculate the fault current:

$$I_a = \frac{U_o}{Z_s} = \frac{230}{0.96} \ A = 240 \ A$$

Next we need to find the operating time for a 32 A BS 88 fuse carrying 240 A. Examination of {Fig 3.15} shows that operation will take place after 0.2 s.

Finally, we need a value for k. From {Table 5.8} we can read this off as 115.

We now have values for I_a, t and k so we can calculate conductor size.

$$S = \frac{\sqrt{(I_a^2 t)}}{k} = \frac{\sqrt{(240^2 \times 0.2)}}{115} \ mm^2 = 0.93 \ mm^2$$

This result suggests that a 1.0 mm² protective conductor will suffice. However, it may be dangerous to make this assumption because the whole calculation has been based on the resistance of a 2.5 mm² conductor. Let us start again assuming a 1.5 mm² protective conductor.

The new size protective conductor has a resistance of 12.1 mΩ/m, see {Table 5.5}, and with the 4 mm² phase conductor gives a total conductor resistance, allowing for increased temperature, of 0.361 Ω. When added to external loop impedance this gives a total earth-fault loop impedance of 1.061 Ω, and a fault current at 230 V of 217 A. From {Fig 3.15} operating time will be 0.35 s. The value of k will be unchanged at 115.

$$S = \frac{\sqrt{(I_a^2 t)}}{k} = \frac{\sqrt{(217^2 \times 0.35)}}{115} \ mm^2 = 1.12 \ mm^2$$

Thus, a 1.5 mm² protective conductor can be used in this case. Note that if the size had been assessed rather than calculated, the required size would be 4 mm², two sizes larger.

Example 5.3

A 230 V, 30 A ring circuit for socket outlets is 45 m long and is to be wired in 2.5 mm² flat twin p.v.c. insulated and sheathed cable including a 1.5 mm² cpc. The circuit is to be protected by a semi-enclosed (rewirable) fuse to BS 3036, and the earth-fault loop impedance external to the installation has been ascertained to be 0.3 Ω. Verify that the 1.5 mm² cpc enclosed in the sheath is adequate.

First use {Table 5.5} to find the resistance of the phase and cpc conductors. These are 7.4 mΩ/m and 12.1 mΩ/m respectively, so for a 45 m length and allowing for temperature factor of 1.2,

$$R_1 + R_2 = \frac{(7.4 + 12.1) \times 1.2 \times 45}{1000} \ \Omega = \frac{19.5 \times 1.2 \times 45}{1000} \ \Omega = 1.05 \ \Omega$$

$$Z_s = Z_E + \frac{R_1 + R_2}{4} = 0.3 + \frac{1.05}{4} \ \Omega = 0.3 + 0.263 \ \Omega = 0.563 \ \Omega$$

The division by 4 is to allow for the ring nature of the circuit.

$$I_a = \frac{U_o}{Z_s} = \frac{230}{0.563} = 409 \ A$$

We must then use the time/current characteristic of {Fig 3.13} to ascertain an operating time of 0.10 s. From {Table 5.8} the value of k is 115.

$$\text{Then} \quad S = \frac{\sqrt{(I_a^2 t)}}{k} = \frac{\sqrt{(409^2 \times 0.10)}}{115} \ mm^2 = 1.12 \ mm^2$$

Since this value is smaller than the intended value of 1.5 mm², this latter value will be satisfactory.

Example 5.4

A 230 V single-phase circuit is to be wired in p.v.c. insulated single core cables enclosed in plastic conduit. The circuit length is 45 m and the live conductors are 16 mm² in cross-sectional area. The circuit will supply fixed equipment, and is to be protected by a 63 A HBC fuse to BS 88. The earth-fault loop impedance external to the installation has been ascertained to be 0.58 Ω. Calculate a suitable size for the circuit protective conductor.

With the information given this time the approach is somewhat different. We know that the maximum disconnection time for fixed equipment is 5 s, so from the time/current characteristic for the 63 A fuse {Fig 3.15} we can see that the fault current for disconnection will have a minimum value of 280 A.

$$\text{Thus, } Z_s = \frac{U_o}{I_a} = \frac{230}{280} \ \Omega = 0.821 \ \Omega$$

If we deduct the external loop impedance, we come to the resistance of phase and protective conductors.

$$R_1 + R_2 = Z_s - Z = 0.821 - 0.58 \ \Omega = 0.241 \ \Omega$$

Converting this resistance to the combined value of R_1 and R_2 per metre,

$$(R_1 + R_2) \text{ per metre} = \frac{0.241 \times 1000}{45 \times 1.2} \ m\Omega/m = 4.46 \ m\Omega/m$$

Consulting {Table 5.5} we find that the resistance of 16 mm² copper conductor is 1.15 mΩ/m, whilst 10 mm² and 6 mm² are 1.83 and 3.08 mΩ/m respectively. Since 1.15 and 3.08 add to 4.23, which is less than 5.13, it would seem that a 6 mm² protective conductor will be large enough. However, to be sure we must check with the adiabatic equation.

$$R_1 + R_2 = \frac{(1.15 + 3.08) \times 1.2 \times 45}{1000} \; \Omega = 0.228 \; \Omega$$

$$Z_s = Z_E + (R_1 + R_2) = 0.58 + 0.228 \; \Omega = 0.808 \; \Omega$$

$$I_a = \frac{U_o}{Z_s} = \frac{230}{0.808} = 285 \; A$$

From {Fig 3.15} the disconnection time for a 63 A fuse carrying 285 A is found to be 3.8 s. From {Table 5.8} the value of k is 115.

$$\text{Then} \quad S = \frac{\sqrt{(I_a^2 t)}}{k} = \frac{\sqrt{(285^2 \times 3.8)}}{115} \; mm^2 = 4.83 \; mm^2$$

Since 4.83 is less than 6.0 then a 6 mm² protective conductor will be large enough to satisfy the requirements.

5.4.6 Unearthed metalwork

If exposed conductive parts are isolated, or shrouded in non-conducing material, or are small so that the area of contact with a human body is limited, it is permissible not to earth them. Examples are overhead line metalwork which is out of reach, steel reinforcing rods within concrete lighting columns, cable clips, nameplates, fixing screws and so on. Where areas are accessible only to skilled or instructed persons, and where ordinary persons are unlikely to enter due to the presence of warning notices, locks and so on, earthing may be replaced by the provision of obstacles which make direct contact unlikely, provided that the installation complies with the Electricity at Work Regulations, 1989.

5.5 EARTH ELECTRODES [542.2]

5.5.1 Why must we have earth electrodes?

The principle of earthing is to consider the general mass of earth as a reference (zero) potential. Thus, everything connected directly to it will be at this zero potential, or above it only by the amount of the volt drop in the connection system (for example, the volt drop in a protective conductor carrying fault current). The purpose of the earth electrode is to connect to the general mass of earth.

With the increasing use of underground supplies and of protective multiple earthing (PME) it is becoming more common for the consumer to be protected with an earth terminal rather than having to make contact with earth using an earth electrode.

5.5.2 Earth electrode types [542.2]

Acceptable electrodes are rods, pipes, mats, tapes, wires, plates, buried or driven into the ground, welded metal reinforcement of concrete (except pre-stressed concrete) buried in the earth and structural steelwork. The pipes of other services such as gas and water must not be used as earth electrodes although they must be bonded to earth as described in {5.4.3}. The sheath and armour of a buried cable may be used with the approval of its owner and provided that arrangements can be made for the person responsible for the installation to be told if the cable is changed, for example, for a type without a metal sheath.

The effectiveness of an earth electrode in making good contact with the general mass of earth depends on factors such as soil type, moisture content, and so on. A permanently-wet situation may provide good contact with earth, but may also limit the life of the electrode since corrosion is likely to be greater. If the ground in which the electrode is placed freezes, there is likely to be an increase in earth resistance. In most parts of the UK an earth electrode resistance in the range 1 Ω to 5 Ω is considered to be acceptable.

The method of measuring the resistance of the earth electrode will be considered in {7.6.1}; the maximum resistance to earth should be no greater than 220 Ω. The earthing conductor and its connection to the earth electrode must be protected from mechanical damage and from corrosion. Accidental disconnection must be avoided by fixing a permanent label as shown in {Fig 5.17} which reads:

SAFETY ELECTRICAL CONNECTION
–
DO NOT REMOVE

Section 2 of Guidance Note 8 provides copious details concerning earth electrodes.

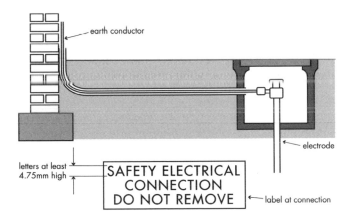

Fig 5.17 Connection of earthing conductor to earth electrode

5.6 PROTECTIVE MULTIPLE EARTHING (PME)
5.6.1 What is protective multiple earthing? (Part 2)
If a continuous metallic earth conductor exists from the star point of the supply transformer to the earthing terminal of the installation, it will run throughout in parallel with the installation neutral, which will be at the same potential. It therefore seems logical that one of these conductors should be removed, with that remaining acting as a combined protective and neutral conductor (PEN). When this is done, we have a TN-C-S installation {5.2.4}. The combined earth and neutral system will apply only to the supply, and not to the installation. Thus, the PME system is the concern of the Supply Company rather than the electrician, although there are special PME requirements to be met by the electrician.

Because of possible dangers with the system which will be explained in the following sub-sections, PME can be installed by the Electricity Supply Company only after the supply system and the installations it feeds have complied with certain requirements. These special needs will be outlined in {5.6.4}.

The great virtue of the PME system is that neutral is bonded to earth so that a phase to earth fault is automatically a phase to neutral fault. The earth-fault loop impedance will then be low, resulting in a high value of fault current which will operate the protective device quickly. It must be stressed that the neutral and earth conductors are kept quite separate within the installation: the main earthing terminal is bonded to the incoming combined earth and neutral conductor by the Elecricity Supply Company. The difficulty of ensuring that bonding requirements are met on construction sites means that PME supplies must not be used. Electricity Safety, Quality and Continuity (ESQC) Regulations forbid the use of PME supplies to feed caravans and caravan sites, as well as boats and marinas. Special precautions may also be necessary in other locations, such as farms, dairies, building sites and swimming pools; at the very least the installation of an earth electrode connected to the main PME earthing terminal should be considered.

5.6.2 Increased fire risk

As with other systems of earth-fault protection, PME does not prevent a fault occurring, but will ensure that the fault protection device operates quickly when that fault appears. For example, if a fault of 2 Ω resistance occurs in a 230 V circuit protected by a 20 A semi-enclosed fuse in a system with an earth-fault loop impedance of 6 Ω the fault current will be 230/(2 + 6) A = 230/8 A = 28.75 A. The fuse would not blow unless the circuit was already loaded, when load current would add to fault current. If the circuit were fully loaded with a load current of 20 A, total current would be 48.75 A and the fuse would operate after about 18 s. During this time, the power produced in the fault would be:

$$P = I^2R = 28.75^2 \times 2 = 1650 \text{ W or } 1.65 \text{ kW}$$

This could easily start a fire. If, however, the earth-fault loop impedance were 1 Ω current would be 76.7 A and the fuse would blow in about 1.6 s and limit the energy in the fault circuit.

5.6.3 Broken neutral conductor

The neutral of a supply is often common to a large number of installations. In the (unlikely) event of a broken neutral, all the conductors on the load side of the break could have a combined neutral and earth potential of the same level as the phase system (230 V to earth). This situation could be very dangerous, because all earthed metalwork would be at 230 V above the potential of the general mass of earth.

To prevent such an event, the Electricity Supply Company connects its combined neutral and earth conductor to earth electrodes at frequent intervals along its run. Whilst this does not entirely remove the danger, it is much reduced. For example, the assumed earth resistance values of {Fig 5.18} show that the maximum possible potential to earth in this case would be 92 V. In practice, much lower resistance values for the earth connections will reduce this voltage. The Electricity Supply Company will ensure the integrity of its neutral conductor.

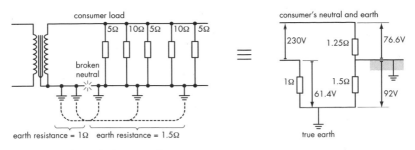

Fig 5.18 Danger due to broken neutral in a PME system

5.6.4 Special requirements for PME-fed installations

[411, 544, 711, 740]

An installation connected to a protective multiple earth supply is subject to special requirements concerning the size of earthing and bonding leads, which are generally larger in cross-section than those for installations fed by supplies with other types of earthing. Full discussions with the Electricity Supply Company are necessary before completing such an installation to ensure that their needs will be satisfied. The cross-sectional area of the equipotential bonding conductor is related to that of the neutral conductor as shown in {Table 5.9}.

Table 5.9 Minimum cross-sectional area of main of main equipotential bonding conductor (From [Table 54.8] of BS 7671:2008)

Neutral conductor c.s.a. (mm²)	Main equipotential bonding conductor c.s.a. (mm²)
35 or less	10
over 35 and up to 50	16
over 50 and up to 95	25
over 95 and up to 150	35
over 150	50

Danger can arise when the non-current carrying metalwork of an installation is connected to the neutral, as is the case with a PME-fed system. The earth system is effectively in parallel with the neutral, and will thus share the normal neutral current. This current will not only be that drawn by the installation itself, but may also be part of the neutral current of neighbouring installations.

It follows that the earth system for an installation may carry a significant current (of the order of tens of amperes) even when the main supply to that installation is switched off. This could clearly cause a hazard if a potentially explosive part of an installation, such as a petrol storage tank, were the effective earth electrode for part of the neutral current of a number of installations. For this reason, the Health and Safety Executive has banned the use of PME in supplies for petrol filling stations.

Installations in petrol filling stations are subject to the Dangerous Substances and Explosive Atmospheres Regulations which came into force in 2003. They replace the Highly Flammable Liquids and Liquefied Petroleum Gases Regulations. Guidance on the design, construction, modification and maintenance of petrol filling stations is published by the Institute of Petroleum (IP), ISBN 0 85293 217 0, and should be studied and followed by those undertaking this very specialised work.

It seems likely that compliance with the ESQC Regulations will require an earth electrode to be installed and connected to the main PME terminal. This will provide reliability and security in the unlikely event of loss of PEN continuity.

5.7 EARTHED CONCENTRIC WIRING

5.7.1 What is earthed concentric wiring?

This is the TN-C system {5.2.5} where a combined neutral and earth (PEN) conductor is used throughout the installation as well as for the supply. The PEN conductor is the sheath of a cable and therefore is concentric with (totally surrounds) the phase conductor(s). The system is unusual, but where employed almost invariably uses mineral insulated cable, the metallic copper sheath being the combined neutral and earth conductor.

5.7.2 Requirements for earthed concentric wiring [411, 543, 705]

Earthed concentric wiring may only be used under very special conditions, which usually involve the use of a private transformer supply or a private generating plant. Since there is no separate path for earth currents, it follows that residual current devices (RCDs) will not be effective and therefore must not be used. The cross-sectional area of the sheath (neutral and earth conductor) of a cable used in such a system must never be less than 4 mm² copper, or 16 mm² aluminium or less than the inner core for a single core cable. All multicore copper mineral insulated cables comply with this requirement, even a 1 mm² two-core cable having the necessary sheath cross-sectional area. However, only single core cables of 6 mm² and below may be used. The combined protective and neutral conductors (sheaths) of such cables must not serve more than one final circuit.

Wherever a joint becomes necessary in the PEN conductor, the contact through the normal sealing pot and gland is insufficient; an extra earth tail must be used as shown in {Fig 5.19}. If it becomes necessary to separate the neutral and protective conductors at any point in an installation, they must not be connected together again beyond that point.

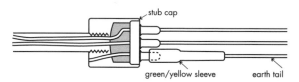

Fig 5.19 Earth tail for use in earthed concentric wiring

5.8 OTHER PROTECTION METHODS

Most of this chapter has been concerned with the protection of the user of the electrical installation against severe shock. The approach has not been to prevent shock altogether, because this is impossible, but to limit its severity by ensuring that the shock current level is low and that the duration of its flow is very short. This section deals with the prevention of shock altogether by the use of special methods.

5.8.1 Class II equipment [412, 413]

Class II equipment has reinforced or double insulation. As well as the basic insulation for live parts, there is a second layer of insulation, either to prevent contact with exposed conductive parts or to make sure that there can never be any contact between such parts and live parts. The outer case of the equipment need not be made of insulating material; if protected by double insulation, a metal case will not present any danger. It must never be connected with earth, so connecting leads are two-core, having no protective conductor. The symbol for a double-insulated appliance is shown in {Fig 5.20}.

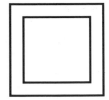

Fig 5.20 British Standard symbol for double insulation

To make sure that the double insulation is not impaired, conducting parts such as metal screws must not pierce it. Nor must insulating screws be used, because there is the possibility that they will be lost and be replaced by metal screws. Any holes in the enclosure of a double-insulated appliance, such as those to allow ventilation, must be so small that fingers cannot reach live parts (IP2X protection). Class II equipment must be installed and fixed so that the double insulation will never be impaired, and so that metalwork of the equipment does not come into contact with the protective system of the main installation. Where the whole of an installation is comprised of Class II equipment, so that there is no protective system installed, the situation must be under proper supervision to make sure that no changes are made which will introduce earthed parts.

5.8.2 Non-conducting location [410, 418, 612]
The non-conducting location is a special arrangement where there is no earthing or protective system because:
1 there is nothing which needs to be earthed
2 exposed conductive parts are arranged so that it is impossible to touch two of them, or an exposed conducting part and an extraneous conductive part, at the same time. The distance between the parts must be at least 2 m, or 1.25 m if they are out of arm's reach. An alternative is to erect suitable obstacles, or to insulate the extraneous conductive parts.

Examples of extraneous conductive parts are water and gas pipes, structural steel-work, and even floors and walls which are not covered with insulating material. Insulation tests on floors and walls are considered in {7.5.2}. There must be no socket outlets with earthing contacts in a non-conducting location. This type of installation could cause danger if earthed metal were introduced in the form of a portable appliance fed by a lead from outside the location.

The potential reached by exposed metalwork within the situation is of no importance because it is never possible to touch two pieces of metalwork with differing voltage levels at the same time. Care must be taken, however, to make sure that a possible high potential cannot be transmitted outside the situation by the subsequent installation of a conductor such as a water or gas pipe. A notice must be erected to state that a non-conducting location exists, and giving details of the person in charge who alone will authorise any work to be undertaken in, or will authorise any equipment to be taken into, the location. If two faults to exposed conductive parts occur from conductors at different potentials (such as a phase and a neutral) and there is a defective bonding system, dangerous potential differences could occur between exposed conductive parts. To prevent this possibility, double pole fuses or circuit breakers must provide overload protection in non-conducting locations.

Non-conducting locations are unusual, and their use must be limited to situations where there is continuous and proper supervision to ensure that the requirements are fully met and are properly maintained. This type of installation should only be considered after consulting a fully qualified electrical engineer.

5.8.3 Earth-free bonding [418.2]

The Regulations permit the provision of an area in which all exposed metal parts are connected together, but not to earth. Inside the area, there can be no danger, even if the voltage to earth is very high, because all metalwork which can be touched will be at the same potential. Care is necessary, hower, to prevent danger to people entering or leaving the area, because then they may be in contact with parts which are inside and others which are outside the area, and hence at differing potentials. A notice must be erected to warn that the bonding conductors in the system must not be connected to earth, and that earthed equipment must not be brought into the situation.

As in the case of non-conducting locations, this type of installation is unusual, and must only be undertaken when designed and specified by a fully qualified electrical engineer. The area inside the protected system is often referred to as a 'Faraday cage'.

5.8.4 Electrical separation [418, 612, 753]

Separating a system completely from others so that there is no complete circuit through which shock current could flow can sometimes ensure safety from shock. It follows that the circuit must be small to ensure that earth impedances are very high and do not offer a path for shock current (see {Fig 5.3(b)}). The source of supply for such a circuit could be a battery or a generating set, but is far more likely to be an isolating transformer with a secondary winding providing no more than 500 V. Such a transformer must comply with BS EN 60742, having a screen between its windings and a secondary winding which has no connection to earth.

There must be no connection to earth and precautions must be taken to ensure, as far as possible, that earth faults will not occur. Such precautions would include the use of flexible cords without metallic sheaths, using double insulation, making sure that flexible cords are visible throughout their length of run, and so on. Perhaps the most common example of a separated circuit is the bathroom transformer unit feeding an electric shaver. By breaking the link to the earthed supply system using the double wound transformer, there is no path to earth for shock current (see {Fig 5.21}).

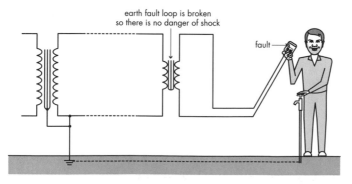

earth fault loop is broken
so there is no danger of shock

fault

Fig 5.21 Bathroom shaver socket to BS EN 60742

5.9 RESIDUAL CURRENT DEVICES (RCDS)

5.9.1 Why do we need residual current devices?

{5.3} has stressed that the standard method of protection is to make sure that an earth fault results in a fault current high enough to operate the protective device quickly so that fatal shock is prevented. However, there are cases where the impedance of the earth-fault loop, or the impedance of the fault itself, are too high to enable enough fault current to flow. In such a case, either:

1 current will continue to flow to earth, perhaps generating enough heat to start a fire, or

2 metalwork which is open to touch may be at a high potential relative to earth, resulting in severe shock danger.

Either or both of these possibilities can be removed by the installation of a residual current device (RCD). In recent years there has been an enormous increase in the use of initials for residual current devices of all kinds. The following list, which is not exhaustive, may be helpful to readers:

RCD   residual current device
RCCD   residual current operated circuit breaker
SRCD   socket outlet incorporating an RCD
PRCD   portable RCD, usually an RCD incorporated into a plug
RCBO   an RCCD which includes over-current protection
SRCBO a socket outlet incorporating an RCBO
CBR   a circuit breaker incorporating residual current protection
RCM   a residual current monitor without tripping or protection capabilities

5.9.2 The principle of the residual current device [514.12]

The RCD is a circuit breaker that continuously compares the current in the phase with that in the neutral. The difference between the two (the residual current) will be flowing to earth, because it has left the supply through the phase and has not returned in the neutral (see {Fig 5.22}). There will always be some residual current in the insulation resistance and capacitance to earth, but in a healthy circuit such current will be low, seldom exceeding 2 mA.

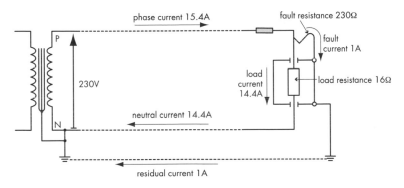

Fig 5.22 The meaning of the term residual current

The purpose of the residual current device is to monitor the residual current and to switch off the circuit quickly if it rises to a dangerous (preset) level. The arrangement of an RCD is shown in simplified form in {Fig 5.23}. The main contacts are closed against the pressure of a spring, which provides the energy to open them when the device trips. Phase and neutral currents pass through identical coils wound in opposing directions on a magnetic circuit, so that each coil will provide equal but opposing numbers of ampere turns when there is no residual current. The opposing ampere turns will cancel, and no magnetic flux will be set up in the magnetic circuit.

Residual earth current passes to the circuit through the phase coil but returns through the earth path, thus avoiding the neutral coil, which will therefore carry less current. This means that phase ampere-turns exceed neutral ampere-turns and an alternating magnetic flux results in the core. This flux links with the search coil, which is also wound on the magnetic circuit, inducing an e.m.f. into it. The value of this e.m.f. depends on the residual current, so it will drive a current to the tripping system depending on the difference between phase and neutral currents. When the amount of residual current, and hence of tripping current, reaches a pre-determined level, the circuit breaker trips, opening the main contacts and interrupting the circuit.

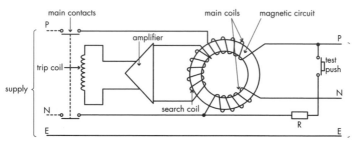

Fig 5.23 Residual current circuit breaker

For circuit breakers operating at low residual current values, an amplifier may be used in the trip circuit. Since the sum of the currents in the phases and neutral of a three-phase supply is always balanced, the system can be used just as effectively with three-phase supplies. In high current circuits, it is more usual for the phase and neutral conductors to simply pass through the magnetic core instead of round coils wound on it.

Operation depends on a mechanical system, which could possibly become stiff when old or dirty. Thus, regular testing is needed, and the RCD is provided with a test button that provides the rated level of residual current to ensure that the circuit breaker will operate. All RCDs are required to display a notice to draw attention to the need for frequent testing which can be carried out by the user, who presses a test button, usually marked T. {Table 5.10} shows the required notice.

Table 5.10 Periodic test notice for residual current device

This installation, or part of it, is protected by a device which automatically switches off the supply if an earth fault develops. Test quarterly by pressing the button marked 'T' or 'Test'. The device should switch off the supply, and should then be switched on to restore the supply. If the device does not switch off the supply when the button is pressed, seek expert advice.

The test circuit is shown in {Fig 5.23}, and provides extra current in the phase coil when the test button is pressed. This extra current is determined by the value of the resistor R.

There are currently four basic types of RCD. Class AC devices are used where the residual current is sinusoidal - this is the normal type which is in the most wide use. Class A types are used where the residual current is sinusoidal and/or includes pulsating direct currents - this type is applied in special situations where electronic equipment is used. Class B is for specialist operation on pure direct current or on impulse direct or alternating current. Class S RCDs have a built-in time delay to provide selectivity (discrimination) (see below).

It must be understood that the residual current is the difference between phase and neutral currents, and that the current breaking ability of the main contacts is not related to the residual operating current value. There is a widely held misunderstanding of this point, many people thinking that the real current setting is the current breaking capability of the device. It is very likely that a device with a breaking capacity of 100 A may have a residual operating current of only 30 mA.

There are cases where more than one residual current device is used in an installation; for example, a complete installation may be protected by an RCD rated at 100 mA whilst a socket intended for equipment outdoors may be protected by a 30 mA device. Selectivity (previously known as discrimination of the two devices then becomes important. For example, if an earth fault giving an earth current of 250 mA develops on the equipment fed by the outdoor socket, both RCDs will carry this fault current, and both will become unbalanced. Since the fault is higher than the operating current of both devices, both will have their trip systems activated. It does not follow that the device with the small operating current will open first, so it is quite likely that the 100 mA device will operate, cutting off the supply to the complete installation even though the fault was on a small part of it. This is a lack of selectivity between the residual current devices. To ensure proper selectivity, the device with the larger operating current has a deliberate delay built into its operation. It is called a time delayed RCD.

5.9.3 Regulations for residual current devices [411, 522, 531, 701, 702, 705, 708, 709, 740, 753]

The primary purpose of the residual current device is to limit the severity of shock in the event of a fault. In other words, it will detect and clear earth faults which otherwise would or could lead to dangerous potential differences between pieces of metalwork which are open to touch. If the sensitivity of the device (its operating residual current) is low enough, it may also be used to limit the shock received from direct contact with a live conductor (basic protection) in the case of the failure of other measures. A problem that may occur here is nuisance tripping, because the operating current may be so low that normal leakage current will cause operation. For example an RCD with a sensitivity of 2 mA will switch off the supply as soon as a shock current of 2 mA flows, virtually preventing a fatal shock. The difficulty is that normal insulation resistance leakage and stray capacitance currents can easily reach this value in a perfectly healthy system, and it may thus be impossible to keep the circuit breaker closed. The total leakage of various items of equipment protected by the RCD should never be so large as to cause unwanted operation. Normal earth leakage current from equipment and appliances will, of course vary with the condition of the device. Maximum permitted leakage currents are listed in Appendix L of the 2nd Edition of Guidance Note 1, and vary from 0.25 mA for Class II appliances to 3.5 mA for information technology equipment.

Some RCDs (usually electronic types) will not switch off unless the mains supply is available to provide power for their operation. In such a case, mains failure may prevent tripping whilst danger is still present, (due to, for example, charged capacitors). Such RCDs may only be used where there is another means of protection under fault conditions, or where the only people using the installation are skilled or instructed so that they are aware of the risk.

In some cases RCDs are designed so that their operating parameters, such as the rated residual current or the time delay, can be adjusted. If such an RCD can be operated by an ordinary person (rather than by a skilled or instructed person) then such adjustments must only be possible by a deliberate act using a key or a tool which results in a visible indication of the setting.

If a residual current circuit breaker is set at a very low sensitivity, it can prevent death from electric shock entirely. However, the problem is that a safe current cannot be determined, because it will vary from person to person, and also with the time for which it is applied. The Regulations require a sensitivity of 30 mA for RCDs intended to provide additional protection from direct contact with a lethal voltage.

An RCD must not be used in an installation with neutral and earth combined (TN-C system using a PEN conductor) because there will be no residual current in the event of a fault to cause the device to operate, since there is no separate path for earth fault currents.

RCD protection is required for all socket outlets rated at up to 32 A. This means in practice that most socket outlet circuits in houses and industrial and commercial situations need RCD protection.

Protection by an RCD with a rating of 30 mA is required for fixed electrical equipment installed in a bathroom or in a swimming pool or sauna room.

Although residual current devices are current-operated, there are circumstances where the combination of operating current and high earth-fault loop impedance could result in the earthed metalwork rising to a dangerously high potential. The Regulations draw attention to the fact that if the product of operating current (A) and earth-fault loop impedance (Ω) exceeds 50, the potential of the earthed metalwork will be more than 50 V above earth potential and hence dangerous. This situation must not be allowed to arise (see {Fig 5.24}).

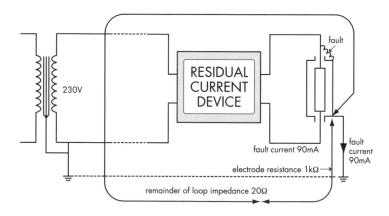

Fig 5.24 Danger with an RCD where fault loop impedance is high. In this case, p.d. from earth to exposed conductive parts will be 1000 Ω x 0.09 A = 90 V,

RCDs must be tested to ensure correct operation within the required operating times. Such tests will be considered in {7.6.3}.

Special requirements apply to RCDs used to protect equipment having normally high earth leakage currents, such as data processing and other computer-based devices. These installations are considered in {8 .8.2}.

5.9.4 Fault voltage operated circuit breakers

These circuit breakers, commonly known in the trade as 'voltage ELCBs', were deleted from the 15th Edition of the Wiring Regulations in 1985, and should not be installed. It is essential that installations where they are still in use should be carefully tested prior to a change to residual current device protection.

5.10 COMBINED FUNCTIONAL AND PROTECTIVE EARTHING [542, 543]

The previous section has made it clear that high earth leakage currents can cause difficulties in protection. The increasing use of data processing equipment such as computers has led to the need for filters to protect against transients in the installation, which could otherwise result in the loss of valuable data. Such filters usually include capacitors connected between live conductors and earth. This has led to large increases in normal earth currents in such installations, and to the need for special regulations for them. Since these are special situations, they will be considered in detail in {8.8}.

Electrical disturbances on the earth system (known as 'earth noise') may cause malfunctions of computer-based systems, and 'clean' mains supplies and earth systems may be necessary. A separate earthing system may be useful in such a case provided that:

1 the computer system has all accessible conductive parts earthed,
2 the main earthing terminal of the computer earth system is connected directly to the main earthing terminal,
3 all extraneous conductive parts within reach of the computer system are earthed to the main earthing terminal and not to the separate computer earth.

Supplementary bonding between the computer earth system and extraneous conductive parts is not necessary.

CHAPTER SIX

Circuits

6.1 BASIC REQUIREMENTS FOR CIRCUITS [132.10]

The Regulations require that installations should be divided into circuits, the purposes being:

1 to prevent danger in the event of a fault by ensuring that the fault current is no greater than necessary to operate the protective system. For example, a large three-phase motor must be connected to a single circuit because the load cannot be subdivided. If, however, a load consisted of three hundred lamps, each rated at 100 W, it would be foolish to consider putting all of this load onto a single circuit. In the event of a fault, the whole of the lighting would be lost, and the fault current needed to operate the protective device (single-phase circuit current would be more than 130 A at 230 V) would be high enough to cause a fire danger at the outlet where the fault occurred. The correct approach would be to divide the load into smaller circuits, each feeding, perhaps, ten lamps.

2 to enable part of an installation to be switched off for maintenance or for testing without affecting the rest of the system.

3 to prevent a fault on one circuit from resulting in the loss of the complete installation (see {3.8.6} on the subject of selectivity or discrimination).

4 to reduce the possibility of unwanted tripping of RCDs (see {5.9}).

5 to reduce the effects of electromagnetic interference (EMI).

6 to reduce the possibility of an isolated circuit being inadvertently energised.

The number of final circuits will depend on the types of load supplied, and must be designed to comply with the requirements for over-current protection, switching and the current-carrying capacity of conductors. Every circuit must be separate from others and must be connected to its own over-current protective fuse or circuit breaker in a switch fuse, distribution board, consumer's unit, etc. See {Figs 6.1 and 6.2}.

A durable notice giving details of all the circuits fed is required to be posted in or near each distribution board. The data required is the equipment served by each circuit, its rating, its design current and its breaking capacity. When the occupancy of the premises changes, the new occupier must be provided with full details of the installation (see reference to the Operating Manual in {7.8.1}). These data must always be kept up to date.

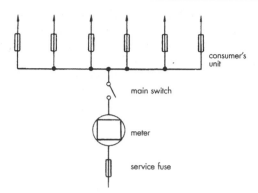

Fig 6.1 Typical arrangements for feeding final circuits in a domestic installation.

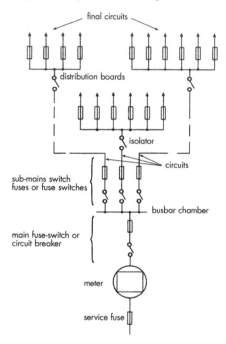

Fig 6.2 An arrangement for main and final circuits in a large installation.

6.2 MAXIMUM DEMAND AND DIVERSITY

6.2.1 Maximum demand [311]

Maximum demand (often referred to as MD) is the largest current normally carried by circuits, switches and protective devices; it does not include the levels of current flowing under overload or short circuit conditions. Assessment of maximum demand is sometimes straightforward. For example, the maximum demand of a 230 V single-phase 8 kW shower heater can be calculated by dividing the power (8 kW) by the voltage (230 V) to give a current of 34.8 A. This calculation assumes a power factor of unity, which is a reasonable assumption for such a purely resistive load.

There are times, however, when assessment of maximum demand is less obvious. For example, if a ring circuit feeds fifteen 13 A sockets, the maximum demand clearly should not be 15 x 13 = 195 A, if only because the circuit protection will not be rated at more than 32 A. Some 13 A sockets may feed table lamps with 60 W lamps fitted, whilst others may feed 3 kW washing machines; others again may not be loaded at all. Guidance is given in {Table 6.1}.

Lighting circuits pose a special problem when determining MD. Each lampholder must be assumed to carry the current required by the connected load, subject to a minimum loading of 100 W per lampholder (a demand of 0.435 A per lampholder at 230 V). Discharge lamps are particularly difficult to assess, and current cannot be calculated simply by dividing lamp power by supply voltage. The reasons for this are:

1 control gear losses result in additional current,
2 the power factor is usually less than unity so current is greater, and
3 control gear often distorts the waveform of the current so that it contains harmonics which are additional to the fundamental supply current.

So long as the power factor of a discharge lighting circuit is not less than 0.85, the current demand for the circuit can be calculated from:

$$\text{current (A)} = \frac{\text{lamp power (W)} \times 1.8}{\text{supply voltage (V)}}$$

For example, the steady state current demand of a 230 V circuit supplying ten 65 W fluorescent lamps would be:

$$I = \frac{10 \times 65 \times 1.8}{230} \text{ A} = 5.09 \text{ A}$$

Switches for circuits feeding discharge lamps must be rated at twice the current they are required to carry, unless they have been specially constructed to withstand the severe arcing resulting from the switching of such inductive and capacitive loads.

Table 6.1 Current demand of outlets
Based on [Table H1] of Guidance Note 1, 4th Edition, Selection and Erection

Type of outlet	Assumed current demand (A)
Socket outlets	Rated demand
Lighting point	Connected load, with minimum of 100 W
Shaver outlet, bell transformer or any equipment of 45 W or less	May be neglected
Household cooker	10 A + 30% of remainder + 5 A for socket in cooker control unit

When assessing maximum demand, account must be taken of the possible growth in demand during the life of the installation. Apart from indicating that maximum demand must be assessed, the Regulations themselves give little help. Suggestions for the assumed current demand of various types of outlet are shown in {Table 6.1}.

6.2.2 Diversity [311]

A domestic ring circuit typically feeds a large number of 13 A sockets but is usually protected by a fuse or circuit breaker rated at 30 A or 32 A. This means that if sockets were feeding 13 A loads, more than two of them in use at the same time would overload the circuit and it would be disconnected by its protective device.

In practice, the chances of all domestic ring sockets feeding loads taking 13 A is small. Whilst there may be a 3 kW washing machine in the kitchen, a 3 kW heater in the living room and another in the bedroom, the chance of all three being in use at the same time is remote. BS 1363 plugs will overheat when carrying 13 A (a 3 kW load at 230 V) and direct connection via a fused spur box should be considered.

If they are all connected at the same time, this could be seen as a failure of the designer when assessing the installation requirements; the installation should have two ring circuits to feed the parts of the house in question.

Most sockets, then, will feed smaller loads such as table lamps, vacuum cleaner, television or audio machines and so on. The chances of all the sockets being used simultaneously is remote in the extreme provided that the number of sockets (and ring circuits) installed is large enough. The condition that only a few sockets will be in use at the same time, and that the loads they feed will be small is called diversity.

By making allowance for reasonable diversity, the number of circuits and their rating can be reduced, with a consequent financial saving, but without reducing the effectiveness of the installation. However, if diversity is over-estimated, the normal current demands will exceed the ratings of the protective devices, which will disconnect the circuits – not a welcome prospect for the user of the installation! Overheating may also result from overloading which exceeds the rating of the protective device, but does not reach its operating current in a reasonably short time. The Regulations require that circuit design should prevent the occurrence of small overloads of long duration.

The sensible application of diversity to the design of an installation calls for experience and a detailed knowledge of the intended use of the installation. Future possible increase in load should also be taken into account. Diversity relies on a number of factors that can only be properly assessed in the light of detailed knowledge of the type of installation, the industrial process concerned where this applies, and the habits and practices of the users. Perhaps a glimpse into a crystal ball to foresee the future could also be useful!

6.2.3 Applied diversity
Apart from indicating that diversity and maximum demand must be assessed, the Regulations themselves give little help. Suggestions of values for the allowances for diversity based on Guidance Note 1 are given in {Table 6.2}.

Distribution boards must not have diversity applied so that they can carry the total load connected to them.

Example 6.1
A shop has the following single-phase loads, which are balanced as evenly as possible across the 400 V three-phase supply.

2 x 6 kW and 7 x 3kW thermostatically controlled water heaters
2 x 3 kW instantaneous water heaters
2 x 6 kW and 1 x 4 kW cookers
12 kW of discharge lighting (sum of tube ratings)
8 x 30 A ring circuits feeding 13 A sockets.

Calculate the total demand of the system, assuming that diversity can be applied.
Calculations will be based on {Table 6.2}.

The single-phase voltage for a 400 V three-phase system is $400/\sqrt{3} = 230$ V.

Table 6.2 Allowance for diversity
Based on [Table H2] of Guidance Note 1, 4th Edition, Selection and Erection

Note the following abbreviations:
X is the full load current of the largest appliance or circuit
Y is the full load current of the second largest appliance or circuit
Z is the full load current of the remaining appliances or circuits

Type of final circuit	Households	Small shops,stores, offices	Hotels, guest houses
		Types of premises	
Lighting	66% total demand	90% total demand	75% total demand
Heating and power	100% up to 10 A + 50% balance	100%X + 75% (Y + Z)	100%X + 80%Y + 60%Z
Cookers	10 A + 30% balance + 5A for socket	100%X + 80%Y + 60%Z	100%X + 80%Y + 60%Z
Motors (but not lifts)		100%X + 80%Y + 60%Z	100%X + 50%(Y+Z)
Instantaneous water heaters	100%X + 100%Y + 25%Z	100%X + 100%Y + 25%Z	100%X + 100%Y + 25%Z
Thermostatic water heaters	100%	100%	100%
Floor warming installations	100%	100%	100%
Thermal storage heating	100%	100%	100%
Standard circuits	100%X + 40%(Y+Z)	100%X + 50%(Y+Z)	100%X +50%(Y+Z)
Sockets and stationary equip.	100%X + 40 %(Y+ Z)	100 %X + 75%(Y+Z)	100 % X + 75 %Y+ 40%Z

All loads with the exception of the discharge lighting can be assumed to be at unity power factor, so current may be calculated from

$$I = \frac{P}{U}$$

Thus the current per kilowatt will be $\frac{1000}{230}$ A = 4.35 A

Water heaters (thermostatic)

No diversity is allowable, so the total load will be:
(2 × 6) + (7 × 3) kW = 12 + 21 kW = 33 kW
This gives a total single-phase current of I = 33 × 4.35 = 143.6 A

Water heaters (instantaneous)

100% of largest plus 100% of next means that in effect there is no allowable diversity.
Single-phase current thus = 2 × 3 × 4.35 = 26.1 A

Cookers

$$100\% \text{ of largest } = 6 \times 4.35 \text{ A} = 26.1 \text{ A}$$
$$80\% \text{ of second } = \frac{80 \times 6 \times 4.35 \text{ A}}{100} = 20.9 \text{ A}$$
$$60\% \text{ of remainder } = \frac{60 \times 4 \times 4.35 \text{ A}}{100} = 10.4 \text{ A}$$
Total for cookers $= 57.4$ A

Discharge lighting

90% of total which must be increased to allow for power factor and control gear losses.

$$\text{Lighting current } = \frac{12 \times 4.35 \times 1.8 \times 90}{100} = 84.6 \text{ A}$$

Ring circuits

First circuit 100%, so current is 30 A

$$75\% \text{ of remainder } = \frac{7 \times 30 \times 75}{100} = 157.5 \text{ A}$$

Total current demand for ring circuits $= 187.5$ A

Total single phase current demand $= 473.1$A

Since a perfect balance is assumed, three phase line current $= \dfrac{473.1 \text{ A}}{3}$

$$= \textbf{157.7 A}$$

6.3 BS 1363 SOCKET OUTLET CIRCUITS [411.3, 553.1]

The BS 1363 socket is the well known fused 13 A rectangular pin type which has become the standard for domestic and commercial use in the UK (see {Fig 6.3}).

6.3.1 The fused plug

In many situations there is a need for socket outlets to be closely spaced so that they are available to feed appliances and equipment without the need to use long and potentially dangerous leads. For example, the domestic kitchen worktop should be provided with ample sockets to feed the many appliances (deep fat fryer, kettle, sandwich toaster, carving knife, toaster, microwave oven, coffee maker, and so on) that are likely to be used. Similarly, in the living room we need to supply television sets, video recorders, stereo players, table lamps, room heaters, etc. In this case, more outlets will be needed to allow for occasional rearrangement of furniture, that may well obstruct access to some outlets.

If each one of these socket outlets were wired back to the mains position or to a local distribution board, large numbers of circuits and cables would be necessary, with consequent high cost. The alternative is the provision of fewer sockets with the penalties of longer leads and possibly the use of multi-outlet adaptors. Because the ideal situation will have closely-spaced outlets, there is virtually no chance of more than a small proportion of them being in use at the same time, so generous allowance can be made for diversity. Thus, cables and protective devices can safely be smaller in size than would be needed if it were assumed that all outlets were simultaneously fully loaded. Twin and even triple socket outlets are common.

Thus a ring circuit protected by a 30 A or 32 A device may well feed twenty socket outlets. It follows that judgement must be used to make as certain as possible that the total loading will not exceed the protective device rating, or its failure and inconvenience will result. Two basic steps will normally ensure that a ring circuit is not overloaded.

1 Do not feed heavy and steady loads (the domestic immersion heater is the most obvious example) from the ring circuit, but make special provision for them on separate circuits.

2 Make sure that the ring circuit does not feed too great an area. This is usually ensured by limiting a single ring circuit to sockets within an area not greater than one hundred square metres.

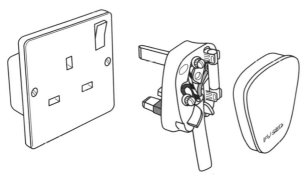

Fig 6.3 Plug and socket to BS 1363

We have already indicated that a 30 A fuse or 32 A circuit breaker is likely to protect a large number of outlets. If this were the only method of protection, there could be a dangerous situation if, for example, a flexible cord with a rating of, say, 5 A developed a fault between cores. {Figure 3.13} shows that a 30 A semi-enclosed fuse will take 5 s to operate when carrying a current of almost 90 A, so the damage to the cord could be extreme. Because of this a further fuse is introduced to protect the appliance and its cord. The fuse is inside the BS 1363 plug, and is rated at 13 A or 3 A, although many other ratings up to 12 A, which are not recognised in the BSS, are available.

A plug to BS 1363 without a fuse is not available. The circuit protection in the distribution board or consumer's unit covers the circuit wiring, whilst the fuse in the plug protects the appliance and its cord as shown in {Fig 6.4}. In this way, each appliance can be protected by a suitable fuse, for example, a 3 A fuse for a table lamp or a 13 A fuse for a 3 kW fan heater.

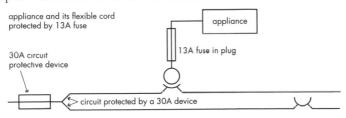

Fig 6.4 Principle of appliance protection by plug fuse

Whilst the installer of the wiring is seldom concerned with the flexible cords of appliances connected to it, he must still offer guidance to users. This will include fitting 3 A fuses in plugs feeding low rated appliances, and the use of flexible cords which are of sufficient cross-section and are as short as possible in the circumstances concerned. Generally, 0.5 mm² cords should be the smallest size connected to plugs fed by 30 A or 32 A ring circuits. Where the cord length must be 10 m or greater, the minimum size should be 0.75 mm² and rubber-insulated cords are preferred to those that are PVC insulated. A common cause of accident was the incorrect connection of plugs by unskilled persons. This has been alleviated by the requirement that manufacturers sell appliances with plugs already fitted.

This type of outlet is not intended for use at high ambient temperatures. A common complaint is the overheating of a fused plug and socket mounted in an airing cupboard to feed an immersion heater; as mentioned above, it is not good practice to connect such a load to a ring circuit, and if unavoidable, final connection should be through a fused spur outlet.

The British fused plug system is probably the biggest stumbling block to the introduction of a common plug for the whole of Europe (the 'Europlug'). The proposed plug is a reversible two-pin type, so would not comply with the Regulations in terms of correct polarity. If we were to adopt it, every plug would need adjacent fuse protection, or would need to be rewired back to its own protective device. In either case, the cost would be very high.

The 17th Edition (BS 7671:2008) requires socket outlets rated at up to 20 A for use by ordinary persons to be protected by 30 mA RCDs. This is a major change from the 16th Edition. The sole exemptions from the RCD protection requirement are

1. socket outlets for use under the supervision of skilled or instructed persons, such as in industrial or commercial situations
2. a specially labelled socket outlet for supplying a particular piece of equipment such as a freezer.

The RCD protection also applies to supplies for mobile equipment used outdoors with a rating not exceeding 32 A.

Where a socket is mounted on a vertical wall, its height above the floor level or the working surface level level must be such that mechanical damage is unlikely. The Building Regulations require that socket outlets and switches should be installed at between 450 mm and 1200 mm above finished floor level. In cases where users may be disabled, a mounting height for socket outlets in the region of 1000 mm is sensible.

6.3.2 The ring final circuit [433.1, 543, 612, Appendix 15]

As stated above, there is a very important change in the 17th Edition. ALL 13 A socket outlets rated at up to 20 A and intended to be used by ordinary persons must now be protected by a 30 mA RCD. The arrangement of a typical ring circuit is shown in {Fig 6.5} and must comply with the following requirements.

1. The floor area served by each ring must not exceed 100 m² for domestic situations. Where ring circuits are used elsewhere (such as in commerce or industry) the diversity must be assessed to ensure that maximum demand will not exceed the rating of the protective device.
2. Consideration should be given to the provision of a separate ring (or radial) circuit in a kitchen. Where there is a washing machine, tumble drier and dishwasher, consideration should be given to the provision of a 4 mm² radial circuit.
3. Where there is more than one ring circuit in the same building, the installed sockets should be shared approximately evenly between them.
4. Cable sizes for standard circuits are as follows:
   a) p.v.c. insulated cable are 2.5 mm² for live (phase and neutral) conductors and 1.5 mm² for the CPC.

b) mineral insulated: 1.5 mm² for all conductors.

These sizes assume that sheathed cables are clipped direct, are embedded in plaster, or have one side in contact with thermally insulating material. Single core cables are assumed to be enclosed in conduit or trunking. No allowance has been made for circuits which are bunched, and the ambient temperature is assumed not to exceed 30°C.

5   The number of unfused spurs fed from the ring circuit must not exceed the number of sockets or fixed appliances connected directly in the ring.

6   Each non-fused spur may feed no more than one single or one twin socket, or no more than one fixed appliance.

7   Fixed loads fed by the ring must be locally protected by a fuse of rating no greater than 13 A or by a circuit breaker of maximum rating 16 A.

8   Fixed equipment such as space heaters, water heaters of capacity greater than 15 litres, and immersion heaters, should not be fed by a ring, but provided with their own circuits.

9   There is no limit to the number of sockets that can be fed from fused spurs to a ring, but see item 6 above.

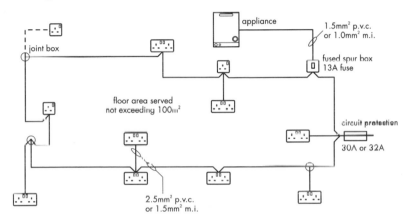

Fig 6.5 Ring circuit feeding socket outlets to BS 1363

6.3.3 The radial circuit [314]

Two types of radial circuit are permitted for socket outlets. In neither case is the number of sockets to be supplied specified, so the number will be subject to the constraints of load and diversity. The two standard circuits are:

1   20 A fuse or 32 A miniature circuit breaker protection with 2.5 mm² live and 1.5 mm² protective conductors (or 1.5 mm² if m.i. cable) feeding a floor area of not more than 50 m². If the circuit feeds a kitchen or utility room, it must be remembered that a 3 kW device such as a washing machine or a tumble dryer takes 13 A at 230 V and that this leaves little capacity for the rest of the sockets.

2   32 A cartridge fuse to BS 88 or miniature circuit breaker feeding through 4 mm² live and 2.5 mm² protective conductors (or 2.5 mm² and 1.5 mm² if m.i. cable) to supply a floor area no greater than 75 m².

The arrangement of the circuits is shown in {Fig 6.6}. 4 mm² may seem to be a large cable size in a circuit feeding 13 A sockets. It must be remembered, however, that the 2.5 mm² ring circuit allows current to be fed both ways round the ring, so that two conductors are effectively in parallel, whereas the 4 mm² cable in a radial circuit must carry all the current.

Radial circuits can be especially economic in a long building where the completion of a ring to the far end could effectively double the length of cable used. As for ring circuits, danger can occur if flexible cords are too small in cross-section, or are too long, or if 3 A fuses are not used where appropriate.

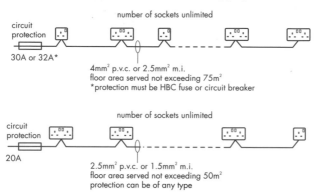

Fig 6.6 Radial circuits for sockets to BS 1363

The minimum cross-sectional area for flexible cords should be:

0.5 mm² where the radial circuit is protected by a 16 A fuse or circuit breaker,
0.75 mm² for a 20 A fuse or circuit breaker,
or 1.0 mm² for a 30 A or 32 A fuse or circuit breaker.

6.4 INDUSTRIAL SOCKET OUTLET CIRCUITS
6.4.1 Introduction
There is no reason at all to prevent the installation of BS 1363 (13 A) socket outlets in industrial situations. Indeed, where light industry, such as electronics manufacture, is concerned, these sockets are most suitable. However, heavy-duty industrial socket outlets are available, and this type is the subject of this section.

6.4.2 BS 196 socket outlet circuits [553.1]
BS 196 sockets are two-pin, non-reversible, with a scraping earth connection. The fusing in the plug can apply to either pole, or the plug may be unfused altogether. Interchangeability is prevented by means of a keyway which may have any of eighteen different positions, identified by capital letters of the alphabet (see {Fig 6.7}). They are available with current ratings of 5 A, 15 A or 30 A.

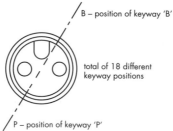

B – position of keyway 'B'

total of 18 different keyway positions

P – position of keyway 'P'

Fig 6.7
Arrangement of socket outlet to BS 196. Two of the eighteen possible key-way positions are shown.

Circuit details for wiring BS 196 outlets can be summarised as:

1 the maximum protective device rating is 32 A
2 all spurs must be protected by a fuse or circuit breaker of rating no larger than 16 A – this means that 30 A outlets cannot be fed from spurs
3 the number of sockets connected to each circuit is unspecified, but proper judgement must be applied to prevent failure of the protective device due to overload
4 cable rating must be no less than that of the protective device for radial circuits, or two thirds of the protective device rating for a ring circuit
5 on normal supplies with an earthed neutral, the phase pole must be fused and the keyway must be positioned at point B (see {Fig 6.7})
6 when the socket is fed at reduced voltage from a transformer with the centre tap on its secondary winding earthed {Fig 6.8}, both poles of the plug must be fused and the keyway must be positioned at P.

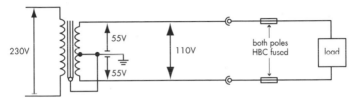

Fig 6.8 Circuit fed from a transformer with a centre-tapped secondary winding.

6.4.3 BS EN 60309-2 socket outlet circuits [553.1]

Plugs and sockets to BS EN 60309-2 are for industrial applications and are rated at 16 A, 32 A, 63 A and 125 A. All but the smallest size must be wired on a separate circuit, but 16 A outlets may be wired in unlimited numbers on radial circuits where diversity can be justified. However, since the maximum rating for the protective device is 20 A, the number of sockets will be small except where loads are very light or where it is certain that few loads will be connected simultaneously. An arrangement of BS EN 60309-2 plugs and sockets is shown in {Fig 6.9}.

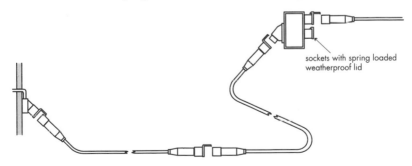

Fig 6.9 Plugs and sockets to BS EN 60309-2

6.5 OTHER CIRCUITS

6.5.1 Lighting circuits [539]

The Building Regulations (Part L) have requirements for electrical installations which are not found in BS 7671. One important example concerns the efficiency of the installed lighting. At least one third of the rooms must be provided with lamps having an efficacy of at least 40 lumens per watt. Whilst fluorescent lamps, compact fluorescent lamps and certain types of discharge lamps usually found in commercial buildings meet this requirement, tungsten and tungsten halogen lamps do not. Thus, bayonet cap (BC) and Edison screw (ES) lamp holders must not be installed in these locations.

Lampholders and ceiling roses must not be used in installations where the supply voltage exceeds 250 V. Lampholders are often mounted within enclosed spaces such as lighting fittings, where the internal temperature may become very high, particularly where filament lamps are used. Care must be taken to ensure that the lampholders, and their associated wiring, are able to withstand the temperature concerned. Where ES lampholders are connected to a system with the neutral at earth potential (TT or TN systems) care must be taken to ensure that the centre contact is connected to the phase conductor and the outer screw to the neutral to reduce the shock danger in the event of touching the outer screw during lamp changing (see {Fig 6.10}).

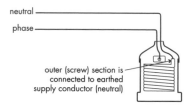

Fig 6.10 Correct connection of ES lampholder

Ceiling roses must not have more than one flexible cord connected to them, and, like the flexible cords themselves, must not be subjected to greater suspended weight than their design permits (see {Table 4.2}). Lampholders in bath or shower rooms must be fitted with a protective shield to prevent contact with the cap whilst changing the lamp (see {Fig 6.11}).

Increasing use is being made of luminaires recessed into the ceiling. These lighting fittings can easily become overheated, especially when incandescent lamps are used and if the space directly above the fitting is provided with insulation intended to prevent the loss of heat. Special covers are available to limit excessive temperatures in such situations.

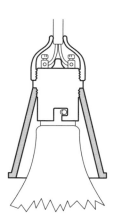

Fig 6.11 Protective shield for a BC lampholder

In large lighting installations, particularly, where fluorescent fittings are involved, consideration should be given to the use of luminaire support couplers (LSCs) or plugs and sockets. Such arrangements facilitate the disconnection of luminaires for electrical maintenance and for cleaning, and may also allow the complete testing of an installation before erection of the luminaires. Many lighting installations are now controlled by sophisticated software (which may switch off the lighting when daylight levels increase or when a room has been unoccupied for a predetermined time). Such devices must be installed to comply with the Regulations.

6.5.2 Cooker circuits [314]

A cooker is regarded as a piece of fixed equipment unless it is a small table-mounted type fed from a plug by a flexible cord. Such equipment must be under the control of a local switch, usually in the form of a cooker control unit. This switch may control two cookers, provided both are within 2 m of it. In many cases this control unit incorporates a socket outlet, although often such a socket is not in the safest position for use to supply portable appliances, whose flexible cords may be burned by the hotplates. It is often considered safer to control the cooker with a switch and to provide a separate socket circuit. The protective device is often the most highly rated in an installation, particularly in a domestic situation, so there is a need to ensure that diversity has been properly calculated (see Table {6.2}).

The diversity applicable to the current demand for a cooker is shown in {Table 6.2} as 10 A plus 30% of the remainder of the total connected load, plus 5 A if the control unit includes a socket outlet. A little thought will show that whilst this calculation will give satisfactory results under most circumstances, there is a danger of triggering the protective device in some cases. For example, at Christmas it is quite likely that two ovens, all four hotplates and a 3 kW kettle could be simultaneously connected. Just imagine the chaos that a disconnected cooker circuit would cause! This alone is a very good reason for being generous with cable and protective ratings.

Example 6.2

A 230 V domestic cooker has the following connected loads:

top oven	1.5 kW
main oven	2.5 kW
grill	2.0 kW
four hotplates	2.0 kW each

The cooker control unit includes a 13 A socket outlet. Calculate a suitable rating for the protective device.

The total cooker load is 1.5 + 2.5 + 2.0 + (4 × 2.0) kW = 14 kW

$$\text{Total current} = \frac{P}{U} = \frac{14000}{230} \text{ A} = 60.9 \text{ A}$$

The demand allowing for diversity is made up of:

the first 10 A	10.0 A

$$+ \text{ 30\% of remainder} = \frac{30 \times (60.9 - 10)}{100} = \frac{30 \times 50.9}{100} = 15.3 \text{ A}$$

+ allowance for socket outlet	5.0 A
Total =	30.3 A

A 32 A protective device is likely to be chosen. The cable rating will depend on correction factors (see {Chapter 5}).

6.5.3 Off-peak appliance circuits

All Electricity Supply Companies continue to offer extremely economic rates for energy taken at off-peak times, usually for seven or ten hours each night (economy 7 or economy 10). The supply meter is usually arranged so that off-peak inexpensive energy can only be obtained from a special pair of terminals, whilst a second pair provide energy throughout the 24 hour period, charging for it at the normal rate. In most cases, energy used at the cheap rate must be stored for use at other times. There are two major methods of storing energy, in both cases involving its conversion to heat.

1 storage heaters, which are used for space heating. Circuits feeding them should always be wired radially, with only one outlet for each. This will help to avoid problems in the event of the storage heater being changed for one of a different power rating. Difficulty in control and higher running costs when compared to gas heating has seen a decline in storage heaters.

2 immersion heaters, the energy being stored as hot water in a lagged tank for use during the day. Since the amount of hot water used is variable, it is usually necessary to have a method of increasing the water temperature should that heated at the cheap rate be used up. This involves the use of a second immersion heater, or a single heater with a double element. Since convection, and hence water heating, takes place mainly above the active heater, an immersion heater placed low in the tank and fed by the off-peak supply will heat the whole tank, whilst a second heater, placed higher in the tank and connected to the normal supply, will be switched on when necessary to top up the temperature of the hot water stored {Fig 6.12(a)}. Sometimes a top mounted dual heater is used for the same purpose as shown in {Fig 6.12(b)}. The normal and off-peak heaters must be supplied through totally separate circuits. 3 kW heaters should be connected permanently to a double pole switch and not fed via a plug and socket.

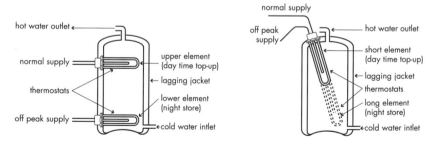

Fig 6.12 Arrangement of immersion heaters for off-peak supplies

6.6 CIRCUIT SEGREGATION

6.6.1 Segregating circuits [528]

There are now only two categories of circuits that may need to be kept separated, which are known as voltage bands.

Voltage Band I is defined as levels of voltage which are too low to provide serious electric shocks; effectively this limits the band to extra-low voltage (ELV), including telecommunications, signalling, bell, control and alarm circuits.

Voltage Band II covers all voltages used in electrical installations not included in Band I. This means that all 230/400 V supplies are included in Band II.

As expected, BS 7671 prohibits Band I and Band II cables sharing the same cable enclosure or multicore cable unless:

1 every cable is insulated for the highest voltage present,
2 each conductor in a multicore cable is insulated for the highest voltage present,
3 they are insulated for their own voltage and installed in separate compartments of a trunking or ducting system,
4 they are installed on a tray with a partition providing separation,
5 a separate conduit or ducting system is provided for each band.

This does mean that BS 7671 allows circuits such as those for fire alarm systems (see {Chapter 9}), emergency lighting, telephones, data transmission, intruder alarms, sound systems, bell and call systems, etc., may now be run together without segregation. BS 5839, on the other hand, makes it clear that fire alarm cables must be separated from all others, and IEE Guidance Note 4 requires that escape lighting cables should be mineral insulated or separated from all others by at least 300 mm. Care must be taken to ensure that circuits are not affected by electrical interference, both electrostatic (due to electric fields) or electro-magnetic (due to electro-magnetic fields).

In some instances it will be necessary for circuit outlets for both voltage bands to share a common box, switch-plate or block. In such a case, the connections of circuits of differing bands must be segregated by a partition, which must be earthed if of metal.

Electrical services must also be kept separate from others that may cause them harm, such as those giving rise to heat, fire, smoke, condensation, etc. Where such separation is impractical, the electrical system must be protected by suitable shielding or by spacing.

6.6.2 Electromagnetic compatibility (EMC) [312, 515]

All electrical equipment must be selected and installed so that it will not affect the supply or cause harmful effects to other equipment. One of the harmful effects is electromagnetic interference (EMI). Whenever current flows in a conductor it sets up a magnetic field; a change in the current will result in a corresponding change in the magnetic field, which will result in the induction of electromotive force (voltage) in any conducting system subject to the field. Whilst induced voltages will usually be very small, they may be considerable when rates of change of current are heavy (for example, circuits feeding lift motors) or when there is a lightning strike in the vicinity. (see also {2.8})

The effects are most pronounced when large metal loops are formed by circuits (perhaps power and data circuits) which are run at a distance from each other but have common earthing and bonding. Power and data cables need to follow common routes to prevent aerial loops which will be subject to induced e.m.f., but with sufficient spacing to prevent interference between them (see {Table 6.4}). The emission standard is BS EN 50081 and the immunity standard BS EN 50082.

The designer needs to consider EMC when planning an installation, and may decide that some of the following measures are appropriate:

1 providing surge protectors and filters for sensitive equipments,
2 proper separation of power and other cables to limit electro-magnetic interference (EMI),

3 using bonding connections which are as short as possible,
4 screening sensitive equipment and bonding of metal enclosures, and
5 avoiding inductive loops by using the same route for cables of different
   systems.
This list is far from exhaustive.

Table 6.4 EMI cable separation distances

Power cable Voltage	Min. separation btw. power & signal cables (m)	Power cable current (A)	Min. separation btw. power & signal cables (m)
115 V	0.25	5	0.24
230 V	0.45	15	0.35
400 V	0.58	50	0.50
3.3 kV	1.10	100	0.60
6.6 kV	1.25	300	0.85
11 kV	1.40	600	1.05

6.6.3 Lift and hoist shaft circuits [528, 56]

A lift shaft may well seem to be an attractive choice for running cables, but this is
not permitted except for circuits which are part of the lift or hoist installation (see
{Fig 6.13}). The cables of the lift installation may be power cables fixed in the shaft
to feed the motor(s), or control cables which feed call buttons, position indicators,
and so on. Trailing cables will feed call buttons, position indicators, lighting and
telephones in the lift itself.

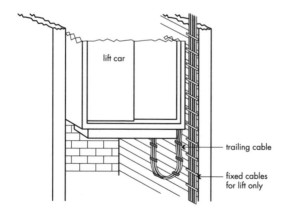

lift car

trailing cable

fixed cables
for lift only

Fig 6.13 Cables installed in lift or hoist shafts

CHAPTER SEVEN

Inspection and testing

7.1 INTRODUCTION

Very useful information concerning inspection and testing will be found in IEE Guidance Note 3, currently in its fourth edition.

7.1.1 The tester

The person who carries out the test and inspection must be competent to do so, and must be able to ensure his own safety, as well as that of others in the vicinity. It follows that he must be skilled and have experience of the type of installation to be inspected and tested so that there will be no accidents during the process to people, to livestock, or to property. The Regulations do not define the term 'competent', but it should be taken to mean a qualified and experienced electrician or electrical engineer.

7.1.2 Why do we need inspection and testing?

There is little point in setting up Regulations to control the way in which electrical installations are designed and installed if it is not verified that they have been followed. For example, the protection of installation users against the danger of fatal electric shock due to fault conditions (previously designated indirect contact) is usually the low impedance of the earth-fault loop; unless this impedance is correctly measured, this safety cannot be confirmed. In this case the test cannot be carried out during installation, because part of the loop is made up of the supply system which is not connected until work is complete. It is important to appreciate that inspection and testing should also take place whilst the installation is being carried out.

In the event of an open circuit in a protective conductor, the whole of the earthed system could become live during the earth-fault loop test. The correct sequence of testing {7.3} would prevent such a danger, but the tester must always be aware of the hazards applying to himself and to others due to his activities. Testing routines must take account of the dangers and be arranged to prevent them. Prominent notices should be displayed to indicate that no attempt should be made to use the installation whilst testing is in progress.

The precautions to be taken by the tester should include the following:

1 make sure that all safety precautions are observed
2 have a clear understanding of the installation, how it is designed and how it has been installed
3 make sure that the instruments to be used for the tests are to the necessary standard (BS EN 61557) and have been recently recalibrated to ensure their accuracy

4 check that the test leads to be used are in good order, with no cracked or broken insulation or connectors, and are fused where necessary to comply with the Health and Safety Executive Guidance Note GS38

5 be aware of the dangers associated with the use of high voltages for insulation testing. For example, cables or capacitors connected in a circuit that has been insulation tested may have become charged to a high potential and may hold it for a significant time.

The dangers associated with earth-fault loop impedance testing have been mentioned in {5.3}.

7.1.3 Information needed by the tester [612]

The installation tester, as well as the user, must be provided with clear indications as to how the installation will carry out its intended purpose.

Table 7.1 Typical schedule of circuits (refer to {Fig 7.1})

Circuit	Fuse (BS 88)	Cable	Feeding
A1	32A	2.5 mm² p.v.c. (X)	ring circuit, printing section
A2	32A	2.5 mm² p.v.c. (X)	ring circuit, drawing office sockets
A3	32A	2.5 mm² p.v.c. (X)	ring circuit, main office sockets
A4	32A	4.0 mm² p.v.c. (X)	store room printer
A5	spare		
A6	spare		
B1	10A	1.5 mm² p.v.c. (Y)	lighting, printing section
B2	10A	1.5 mm² p.v.c. (Y)	lighting, drawing office
B3	10A	1.5 mm² p.v.c. (Y)	lighting, main office
B4	10A	1.5 mm² p.v.c. (Y)	lighting, corridors and toilets
B5	spare		
B6	spare		
C1	32A	2.5 mm² p.v.c. (X)	ring circuit, paper store
C2	32A	2.5 mm² p.v.c. (X)	ring circuit, binding section
C3	spare		
C4	spare		
C5	10A	1.5 mm² p.v.c. (Y)	lighting, paper store
C6	10A	1.5 mm² p.v.c. (Y)	lighting, binding section

To this end, the person carrying out the testing and inspection must be provided with the following data:

1 the type of supply to be connected, i.e. single- or three- phase

2 the assessed maximum demand (see {6.2.1})

3 the earthing arrangements for the installation

4 full details of the installation design, including the number and position of mains gear and of circuits, (see {Tables 7.1 and 7.2, Fig 7.1} for a typical example)

5 all data concerning installation design, including calculation of live and protective conductor sizes, maximum demand, etc.

6 the method chosen to prevent electric shock in the event of an earth fault.

7 a list of all circuits or equipment vulnerable to damage by testing (see {7.2.1 item 3})

Without this complete information the tester cannot verify either that the installation will comply with the Regulations, or that it has been installed in full accordance with the design.

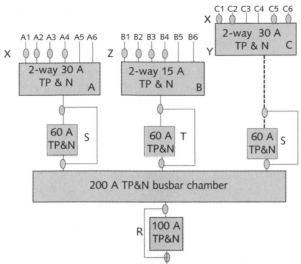

Fig 7.1 Typical mains gear schematic diagram. See also {Tables 7.1 & 7.2}

7.2 INSPECTION

7.2.1 Notices and other identification [514]

The installation tester, as well as the user, must have no difficulty in identifying circuits, fuses, circuit breakers, etc. He must make sure that the installation is properly equipped with labels and notices, which should include:

1 Labels for all fuses and circuit breakers to indicate their ratings and the circuits protected

2 Indication of the purpose of main switches and isolators

3 A diagram or chart at the mains position showing the number of points and the size and type of cables for each circuit, the method of providing protection under normal conditions (previously known as direct contact) and details of any circuit in which there is equipment such as passive infra-red detectors or electronic fluorescent starters vulnerable to the high voltage used for insulation testing.

4 Warning of the presence of voltages exceeding 250 V on an equipment or enclosure where such a voltage would not normally be expected.

5 Warning that voltage exceeding 250 V is present between separate pieces of equipment which are within arm's reach

6 A notice situated at the main intake position to draw attention to the need for periodic testing (see {Table 7.12})

7 A warning of the danger of disconnecting protective conductors (earth wires) at the point of connection of:
a) the earthing conductor to the earth electrode
b) the main earth terminal, where separate from main switchgear
c) bonding conductors to extraneous conductive parts
The notice should read

> **Safety electrical connection – do not remove**

8 A notice to indicate the need for periodic testing of an RCD as indicated in {5.9.2}

9 A notice for caravans so as to draw attention to the connection and disconnection procedure as indicated in {Table 8.9}

10 Warning of the need for operation of two isolation devices to make a piece of equipment safe to work on where this applies

11 A schedule at each distribution board listing the items to be disconnected (such as semiconductors) so that they will not be damaged by testing.

12 A drawing that shows clearly the exact position of all runs of buried cables.

Table 7.2 Typical mains gear schedule (refer to {Fig 7.1})

Item	Situation	Supplying	Cable size	BS 88 fuse size
Dist.board A 2-way 30 A TP&N	Switch room	Ground floor sockets	16 mm² p.v.c (S)	4 x 32 A (2 spare ways)
Dist. board B 2-way 15 A TP&N	Switch room	Ground floor lighting	6 mm² p.v.c. (T)	4 x 10 A (2 spare ways)
Dist. board C 2-way 30 A TP&N	Upper store	Upper floor installation	16 mm² p.v.c. (S)	2 x 32 A 2 x 10 A (2 spare ways)
Board A switch-fuse 60 A TP&N	Switch room	Dist. board A	16 mm² p.v.c. (S)	3 x 60 A
Board B switch-fuse 30 A TP&N	Switch room	Dist. board B	6 mm² p.v.c. (T)	3 x 30 A
Board C switch-fuse 60 A TP&N	Switch room	Dist. board C	16 mm² p.v.c. (S)	3 x 60 A
Main fuse-switch 100 A TP&N	Switch room	Complete installation	70 mm² p.v.c. (R)	3 x 100 A

7.2.2 Inspection [611]

Before testing begins it is important that a full inspection of the complete installation is carried out with the supply disconnected. The word 'inspection' has replaced 'visual inspection', indicating that all the senses (touch, hearing and smell, as well as sight) must be used. Although not directly mentioned in BS 7671, thermal imaging is an excellent method for discovering overheating of parts of the installation. It follows that it can only be used where an installation is still in use, or has only recently been de-energised. In the former case, it is very important to stress the dangers of working with a live electrical system. The main purpose of the inspection is to confirm that the equipment and materials installed:

1 are not obviously damaged or defective so that safety is reduced

2 have been correctly selected and erected

3 comply with the applicable British Standard or the acceptable equivalent

4 are suitable for the prevailing environmental conditions

An inspection check-list is shown in {Table 7.3}. Some inspections are best carried out whilst the work is in progress. A good example is the presence of fire barriers within trunking or around conduit where they pass through walls.

Table 7.3 Inspection check list (from [611.3] of BS 7671: 2008)

1   Identification of conductors and circuits
2   Mechanical protection for cables and routing in safe zones
3   Connection of conductors
4   Correct connection of lampholders, socket outlets, etc.
5   Connection of single-pole switches in phase conductors only
6   Checking of design calculations to ensure that correct live and protective conductors
     have been selected in terms of their current-carrying capacity and volt drop
7   Presence of fire barriers, suitable seals and protection against fire
8   Protection of live parts by insulation to prevent direct contact with live conductors
9   Protection against fault contact by the use of:-
     a) protective conductors
     b) earthing conductors
     c) main and supplementary equipotential bonding conductors
     d) earthing for combined protective and functional purposes
     e) use of Class II equipment or equivalent insulation
     f) non-conducting location and the absence of protective conductors
     g) earth-free local equipotential bonding
     h) electrical separation
     i) SELV
     j) PELV
     k) double insulation
     l) obstacles and placing out of reach
     m) barriers and enclosures
10 Prevention of mutual detrimental influence
11 Appropriateness and siting of switches and isolators
12 Undervoltage protective devices
13 Adequacy of access to switchgear and equipment
14 Labelling of fuses, circuit breakers, circuits, switches and terminals
15 Provision of RCDs, and discrimination between them when necessary
16 Settings and ratings of devices for protection against fault contact and against overcur-
     rent
17 The presence of diagrams, instructions, notices, warnings, etc.
18 Selection of protective measures and equipment in the light of the external influences
     involved (such as the presence of disabled people)
19 Supplies for safety services installed, e.g. fire alarms
20 The presence of a re-test notice (see {Table 7.4}).

7.2.3 Periodic inspection and testing [62]

No electrical installation, no matter how carefully designed and erected, can be expected to last forever. Deterioration will take place due to age as well as due to normal wear and tear. With this in mind, the Regulations require regular inspection and testing to take place so that the installation can be maintained in a good and a safe condition. It is now a requirement of the Regulations that the installation user should be informed of the need for periodic testing, and the date on which the next test is due. A notice, fixed at or near the origin of the installation, must state the required intervals between periodic inspections and tests. BS 7671 gives no specific guidance on the required frequency of periodic inspection and testing, simply indicating that intervals will depend on the type of installation, the frequency and quality of maintenance and the external influences to which it is subjected. {Table 7.4} makes suggestions for periods, although it is stressed that the figures provided are not official. In some cases, such as places of public entertainment, statutory intervals apply and must be carefully followed.

Accessories, switchgear etc should be carefully examined for signs of overheating. Structural changes may have impaired the safety of an installation, as may have changes in the use of space. The use of extension leads must be discouraged, if only because of the relatively high loop impedance they introduce.

It is important to appreciate that the regular inspection and testing of all electrical installations is a requirement of the Electricity at Work Regulations. The time interval concerned will, of course, depend on the type of installation and on the way in which it is used. {Table 7.4} shows the suggested intervals between periodic tests and inspections. Where the installation is under continuous and effective supervision by skilled persons, a suitable programme of continuous monitoring and maintenance can replace periodic tests and inspections.

Table 7.4 Suggested periods between periodic tests and inspections and between routine installation checks From Table 3.2 of Guidance Note 3

Type of installation	Routine check	Maximum period between inspection and testing
Domestic	–	Change of occupancy or 10 years
Commercial	1 year	Change of occupancy or 5 years
Educational	4 months	5 years
Hospitals	1 year	5 years
Industrial	1 year	5 years
Offices	1 year	5 years
Shops	1 year	5 years
Laboratories	1 year	5 years
Cinemas	1 year	3 years *
Churches	1 year	5 years
Leisure complexes (not swimming pools)	1 year	3 years *
Places of public entertainment	1 year	3 years *
Restaurants and hotels	1 year	5 years *
Theatres	1 year	3 years *
Public houses	1 year	5 years *
Village halls and similar	1 year	5 years
Agricultural/horticultural	1 year	3 years
Caravans	1 year	3 years
Caravan parks	6 months	1 year *
Highway power supplies	as convenient	6 years
Marinas	4 months	1 year
Swimming pools	4 months	1 year*
Emergency lighting	daily or monthly	3 years **
Fire alarms	daily or weekly	1 year
Launderettes	1 year	1 year *
Petrol filling stations	1 year	1 year *
Construction sites	3 months	3 months

* Local Authority Conditions of Licence may be different
** Other intervals may be required for testing of batteries and generators.

7.3 TESTING SEQUENCE

7.3.1 Why is correct sequence important?

Testing can be hazardous, both to the tester and to others who are within the area of the installation during the test. The danger is compounded if tests are not carried out in the correct sequence.

For example, it is of great importance that the continuity, and hence the effectiveness, of protective conductors is confirmed before the insulation resistance test is carried out. The high voltage used for insulation testing could appear on all extraneous metalwork associated with the installation in the event of an open-circuit protective conductor if insulation resistance is very low.

Again, an earth-fault loop impedance test cannot be conducted before an installation is connected to the supply, and the danger associated with such a connection before verifying polarity, protective system effectiveness and insulation resistance will be obvious. Any test that fails to produce an acceptable result must be repeated after remedial action has been taken. Any other tests, whose results may have been influenced by the fault in question must also be repeated.

7.3.2 Correct testing sequence [612]

Some tests will be carried out before the supply is connected, whilst others cannot be performed until the installation is energised. {Table 7.5} shows the correct sequence of testing to reduce the possibility of accidents to the minimum.

Table 7.5 Correct sequence for safe testing (From BS 7671:2008, 6.2 to 6.7)

BEFORE CONNECTION OF THE SUPPLY
1   continuity of protective conductors
2   main and supplementary bonding continuity
3   continuity of ring final circuit conductors
4   insulation resistance
5   protection by SELV and PELV
6   site applied insulation
7   protection by electrical separation
8   protection by FELV
9   protection by barriers
10 insulation of non-conducting floors and walls
11 polarity
12 earth electrode resistance if a dedicated tester is used. (see also 13 below)

WITH THE SUPPLY CONNECTED
13 earth electrode resistance if an earth fault loop tester or ammeter and voltmeter are used
14 confirm correct polarity
15 earth fault loop impedance
16 correct operation of residul current devices
17 correct operation of switches and isolators
18 prospective fault current
19 phase sequence
20 verification of volt drop

7.4 CONTINUITY TESTS

7.4.1 Protective conductor continuity [612]

All protective and bonding conductors must be tested to ensure that they are electrically safe and correctly connected. {7.7} gives test instrument requirements. Provided that the supply is not yet connected, it is permissible to disconnect the protective and equipotential conductors from the main earthing terminal to carry out testing. Where the mains supply is connected, as will be the case for periodic testing, the protective and equipotential conductors must not be disconnected because if a fault occurs these conductors may rise to a high potential above earth. In this case, an earth-fault loop tester can be used to verify the integrity of the protective system.

Where earth-fault loop impedance measurement of the installation is carried out, this will remove the need for protective conductor tests because that conductor forms part of the loop. However, the loop test cannot be carried out until the supply is connected, so testing of the protective system is necessary before supply connection, because connection of the supply to an installation with a faulty protective system could lead to danger.

There are three methods for measurement of the resistance of the protective conductor.

1 Using the neutral conductor as a return lead

A temporary link is made at the distribution board between neutral and protective conductor systems. Don't forget to remove the link after testing. The low resistance test instrument is then connected to the earth and neutral of the point from which the measurement is taken (see {Fig 7.2}). This gives the combined resistance of the protective and neutral conductors back to the distribution board. Then

$$R_p = R \times \frac{A_n}{A_n + A_p}$$

where R_p is the resistance of the protective conductor
  R is the resistance reading taken
  A_n is the cross-sectional area of the neutral conductor
  A_p is the cross-sectional area of the protective conductor.

Note that the instrument reading taken in this case is the value of the resistance $R_1 + R_2$ calculated from {Table 5.5} (see {7.4.4}). This method is only valid if both conductors have the same length and both are copper; in most cases where steel conduit or trunking is not used as the protective conductor, the test will give correct results.

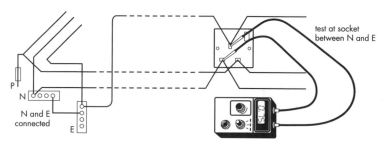

Fig 7.2 Protective conductor continuity test using the neutral conductor as the return lead

2 Using a long return lead

This time a long lead is used which will stretch from the main earthing terminal to every point of the installation.

First, connect the two ends of this lead to the instrument to measure its resistance. Make a note of the value, and then connect one end of the lead to the main earthing terminal and the other end to one of the instrument terminals.

Second, take the meter with its long lead still connected to the point from which continuity measurement is required, and connect the second meter terminal to the protective conductor at that point.

The reading then taken will be the combined resistance of the long lead and the protective conductor, so the protective conductor resistance can be found by subtracting the lead resistance from the reading.

$$R_p = R - R_L$$

where R_p is the resistance of the protective conductor

R is the resistance reading taken

R_L is the resistance of the long lead.

Some modern electronic resistance meters have a facility for storing the lead resistance at the touch of a button, and for subtracting it at a further touch.

3 Where ferrous material forms all or part of the protective conductor

There are some cases where the protective conductor is made up wholly or in part by conduit, trunking, steel wire armour, and the like. The resistance of such materials will always be likely to rise with age due to loose joints and the effects of corrosion. Three tests may be carried out, those listed being of increasing severity as far as the current-carrying capacity of the protective conductor is concerned. They are:

1 A standard ohmmeter test as indicated in 1 or 2 above. This is a low current test which may not show up poor contact effects in the conductor. Following this test, the conductor should be inspected along its length to note if there are any obvious points where problems could occur.

2 If it is felt by the inspector that there may be reasons to question the soundness of the protective conductor, a phase-earth loop impedance test should be carried out with the conductor in question forming part of the loop. This type of test is explained more fully in {7.4.4}

3 If the protective conductor resistance is suspect, the high current test using 1.5 times the circuit design current (with a maximum of 25 A) may be used (see {Fig 7.3}. The protective circuit resistance together with that of the wander lead can be calculated from:

$$\frac{\text{voltmeter reading (V)}}{\text{ammeter reading (A)}}$$

Subtracting wander lead resistance from the calculated value will give the resistance of the protective system.

The resistance between any extraneous conductive part and the main earthing terminal should be 0.05 Ω or less; all supplementary bonds are also required to have the same resistance.

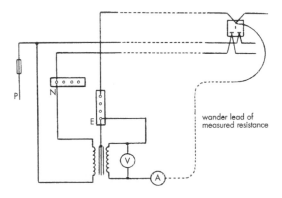

Fig 7.3 High current a.c. test of a protective conductor

7.4.2 Ring final circuit continuity [612.2]

The ring final circuit, feeding 13 A sockets, is extremely widely used, in domestic and in commercial or industrial situations. It is very important that each of the three rings associated with each circuit (phase, neutral and protective conductors) should be continuous and not broken. If this happens, the circuit conductors will not properly share current. {Fig 7.4} shows how this will happen. {Fig 7.4(a)} shows a ring circuit feeding ten socket outlets, each of which is assumed to supply a load taking a current of 3 A. In simple terms, current is then shared between the conductors, so that each could have a minimum current carrying capacity of 15 A. {Fig 7.4(b)} shows the same ring circuit with the same loads, but broken between the ninth and tenth sockets. It can be seen that now one cable will carry only 3 A whilst the other (perhaps with a current rating of 20 A) will carry 27 A. The effect will occur in any broken ring, whether simply one live conductor or both are broken.

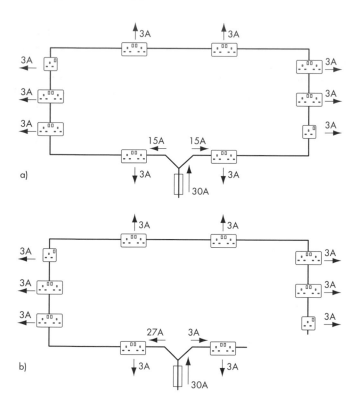

Fig 7.4 Illustrating the danger of a break in a ring final circuit
  a) unbroken ring with correct current sharing
  b) broken ring with incorrect current sharing

It is similarly important that there should be no 'bridge' connection across the circuit. This would happen if, for example, two spurs from different points of the ring were connected together as shown in {Fig 7.5}, and again could result in incorrect load sharing between the ring conductors.

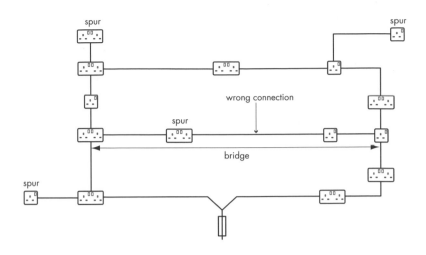

Fig 7.5 A 'bridged' final ring circuit

The tests of the ring final circuit will establish that neither a broken nor a bridged ring has occurred. The following suggested test is based on the Guidance Note 3 on Inspection and Testing issued by the IEE.

Test 1

This test confirms that complete rings exist and that there are no breaks. To complete the test, the two ends of the ring cable are disconnected at the distribution board. The phase conductor of one side of the ring and the neutral from the other (P_1 and N_2) are connected together, and a low resistance ohmmeter used to measure the resistance between the remaining phase and the neutral (P_2 and N_1). {Figure 7.6} shows that this confirms the continuity of the live conductors. To check the continuity of the circuit protective conductor, connect the phase and CPC of different sides together (P_1 and E_2) and measure the resistance between phase and CPC of the other side (P_2 and E_1). The result of this test will be a measurement of the resistance of live and protective conductors round the ring, and if divided by four gives ($R_1 + R_2$) which will conform to the values calculated from {Table 5.5}.

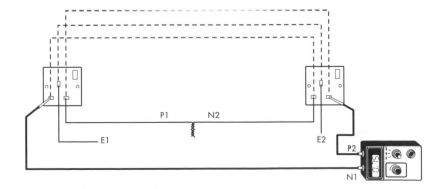

Fig 7.6 Test to confirm the continuity of a ring final circuit

Test 2

This test will confirm the absence of bridges in the ring circuit, see {Fig 7.7}. First, the phase conductor of one side of the ring is connected to the neutral of the other (P_1 and N_2) and the remaining phase and neutral are also connected together (P_2 and N_1). The resistance is then measured between phase and neutral contacts of each socket on the ring. If the results of these measurements are all substantially the same (within 0.05 Ω), the absence of a bridge is confirmed. If the readings are different, this will indicate the presence of a bridge or may be due to incorrect connection of the ends of the ring. If they are connected P_1 to N_1 and P_2 to N_2 then readings will increase or reduce as successive measurements round the ring are taken, as is the case where a bridge exists. Whilst this misconnection is easily avoided when using sheathed cables, a mistake can be made very easily if the system consists of single-core cables in conduit. It may be of interest to note that the resistance reading between phase and neutral outlets at each socket should be one quarter of the phase/neutral reading of Test 1.

Measurements are also taken at each socket on the ring between the phase and the protective conductor with the temporary connection made at the origin of the ring between P_1 to E_2 and between P_2 to E_1. Substantially similar results will indicate the absence of bridges, although sockets wired on spurs will give very slightly higher readings.

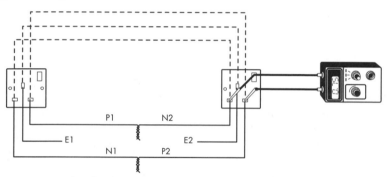

Fig 7.7 Test to confirm the absence of bridges in a ring final circuit

7.4.3 Correct polarity [612.2]

If a single-pole switch or fuse is connected in the neutral of the system rather than in the phase, a very dangerous situation may result as illustrated in {Fig 7.8}.

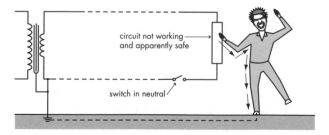

Fig 7.8 The danger in breaking the neutral of a circuit

It is thus of the greatest importance that single-pole switches, fuses and circuit breakers are connected in the phase (non-earthed) conductor, and verification of this connection is the purpose of the polarity test. Also of importance is to test that the outer (screw) connection of ES lampholders is connected to the earthed (neutral) conductor, as well as the outer contact of single contact bayonet cap (BC) lampholders. The test may be carried out with a long wander lead connected to the phase conductors at the distribution board and to one terminal of an ohmmeter or a continuity tester on its low resistance scale. The other connection of the device is equipped with a shorter lead which is connected in turn to switches, centre lampholder contacts, phase sockets of socket outlets and so on. A very low resistance reading indicates correct polarity (see {Fig 7.9}).

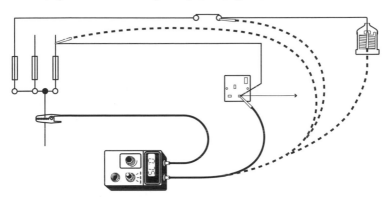

Fig 7.9 Polarity test of an installation

To avoid the use of a long test lead, a temporary connection of phase to protective systems may be made at the mains position A simple resistance test between phase and protective connections at each outlet will then verify polarity. In the unlikely event of the phase and protective conductor connections having been transposed at the outlet, correct polarity will still be shown by this method; this error must be overcome by visual verification.

> **Don't forget to remove the temporary connection afterwards!**

Special care in checking polarity is necessary with periodic tests of installations already connected to the supply, which must be switched off before polarity testing. It is also necessary to confirm correct connection of supply phase and neutral. Should they be transposed, all correctly-connected single-pole devices will be in the neutral, and not in the phase conductor.

One practical method of checking polarity and continuity of ring or radial circuits for socket outlets is to connect two low power lamps to a 13 A plug. One is connected between phase and neutral, and the other between phase and earth. Plugging in at each socket tests correct polarity and the continuity of live and protective conductors when both indicators light. It is important where RCD protection is employed to use very low power indicator lamps such as neon or LED devices (with suitable current limiting resistors where necessary). The smallest filament lamp will take sufficient current from phase to earth to trip most RCDs.

7.4.4 Measurement of $R_1 + R_2$ [612.2]

In {5.3.6} and in {Table 5.5} we considered the value of the resistance of the phase conductor plus that of the protective conductor, collectively known as $(R_1 + R_2)$.

This can be measured with a low resistance reading ohmmeter as described in item 1 of {7.4.1} (strictly the phase, rather than the neutral resistance is being measured here, but there should be no difference) and will be necessarily at the ambient temperature which applies at the time. Multiplication by the correction factor given in {Table 7.6} will adjust the measured resistance to its value at 20°C.

Table 7.6 Correction factors for ambient temperature for $R_1 + R_2$ measurement

Test ambient temperature °C	Correction factor
5	1.06
10	1.04
15	1.02
20	1.00
25	0.98

These will be the resistance values at 20°C and must be adjusted to take account of the increase in resistance of the conductor material due to the increased temperature under normal operating conditions. This second correction factor depends on the ability of the insulation to allow the transmission of heat, and values will be found in {Table 7.7} for three of the more common types of insulating material.

An alternative method for measuring $(R_1 + R_2)$ is to carry out a loop impedance test at the extremity of the final circuit and to deduct the external loop impedance for the installation (Z_E). Strictly it is not correct to add and subtract impedance and resistance values, but any error should be minimal.

Table 7.7 Temperature correction factors for insulation

Insulation type	Correction factor	
	Bunched or as a cable core	Not bunched or sheathed cable
thermoplastic (p.v.c.)	1.20	1.04
85°C rubber	1.26	1.04
90°C thermosetting	1.28	1.04

7.5 INSULATION TESTS

7.5.1 Testing insulation resistance [612.3]

The resistance between phase and neutral conductors, or from live conductors to earth, will result in a leakage current. This current could cause deterioration of the insulation, as well as involving a waste of energy, which would increase the running costs of the installation. Thus, the resistance between poles or to earth must never be less than one megohm (1.0 MΩ) for the usual supply voltages. Note that this minimum value has been increased from 0.5 MΩ as required by the 16th Edition of BS 7671. In addition to the leakage current due to insulation resistance, there is a further current leakage in the reactance of the insulation, because it acts as the dielectric of a capacitor. This current dissipates no energy and is not harmful, but we wish to measure the resistance, not the reactance, of the insulation, so a direct voltage is used to prevent reactance from being included in the measurement. Insulation will sometimes have high resistance when low potential differences apply across it, but will break down and offer low resistance when a higher voltage is applied. For this reason, the high levels of test voltage shown in {Table 7.8} are necessary. {7.7.1} gives test instrument requirements.

Before commencing the test it is important that:

1 electronic equipment that could be damaged by the application of the high test voltage should be disconnected. Included in this category are electronic fluorescent starter switches, touch switches, dimmer switches, LEDs, power controllers, delay timers, switches associated with passive infra-red detectors (PIRs), RCDs with electronic operation etc. An alternative to disconnection is to ensure that phase and neutral are connected together before an insulation test is made between them and earth.

2 capacitors and indicator or pilot lamps must be disconnected or an inaccurate test reading will result.

Table 7.8 Required test voltages and minimum insulation resistance
(from [Table 61] of BS 7671: 2008)

Nominal circuit voltage (V)	Test voltage (V)	Minimum insulation resistance (MΩ)
Extra-low voltage circuits supplied from a safety isolating transformer	250	0.5
Up to 500 V except for above	500	1.0
Above 500 V up to 1000 V	1000	1.0

The insulation resistance tester must be capable of maintaining the required voltage when providing a steady state current of 1 mA.

Where any equipment is disconnected for testing purposes, it must be subjected to its own insulation test, using 250 V, but the measured insulation resistance must be not less than 1.0 MΩ. The result must conform with that specified in the British Standard concerned, or be at least 0.5 MΩ if there is no Standard.

The earth tests must be carried out on the complete installation with the main switch on, with phase and neutral connected together, with lamps and other equipment disconnected, but with fuses in, circuit breakers closed and all circuit switches closed. Where two-way switching is wired, only one of the two strapper wires will be tested. To test the other, both two-way switches should be operated and the system retested. If desired, the installation can be tested as a whole, when a value of at least 1.0 MΩ should be achieved, see {Fig 7.10}. In the case of a very large installation where there are many earth paths in parallel, the reading would be expected to be lower. If this happens, the installation should be subdivided and retested, when each part must meet the minimum requirement.

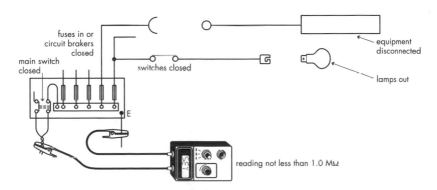

Fig 7.10 Insulation test to earth

The tests to earth {Fig 7.10} and between poles {Fig 7.11} must be carried out as indicated, with a minimum acceptable value for each test of 1.0 MΩ. However, where a reading of less than 2 MΩ is recorded for an individual circuit, (the minimum value required by the Health and Safety Executive), there is the possibility of defective insulation, and remedial work may be necessary. A test result of 2 MΩ may sometimes be unsatisfactory. If such a reading is the result of a re-test, it is necessary to consult the data from previous tests to identify deterioration. A visual inspection of cables to determine their condition is necessary during periodic tests; perished insulation may not always give low insulation readings. As indicated above, tests on SELV and PELV circuits are carried out at 250 V. However tests between these circuits and the live conductors of other circuits must be made at 500 V. Tests to earth for PELV circuits are at 250 V, whilst FELV circuits are tested as LV circuits at 500 V. Readings of less than 5 MΩ will require further investigation.

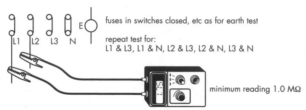

Fig 7.11 Insulation test between poles

7.5.2 Tests of non-conducting floors and walls [418.1, 612.5, Appendix 13]
These tests apply to special non-conducting locations that are under the supervision of skilled or instructed persons. The requirements are

1 there must be no protective conductors
2 if socket outlets are used they must not have an earthing contact
3 it should be impossible for any person to touch two exposed conductive parts at the same time
4 floors and walls must be insulating.

To test this last item and so to make sure that the floors and walls are non-conducting, their insulation has to be tested. These are unusual locations and the electrician is recommended to consult the Regulation references given above if required to carry out work of this nature. The requirements are shown in {Fig 7.12}.

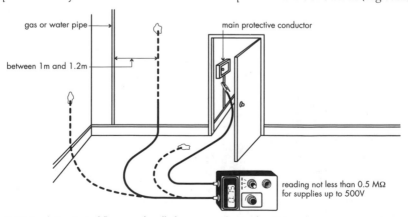

Fig 7.12 Insulation test of floors and walls for non-conducting location

7.5.3 Tests of barriers and enclosures [612.4.5]

Throughout the Regulations reference is made to the use of barriers and enclosures to prevent contact with live parts, basic protection (formerly known as direct contact). If manufactured equipments comply with the British Standards concerned, they will not need further testing, but where barriers and enclosures have been provided during erection of the installation, they must be tested. Full details of the IP classification system will be found in {Table 2.1}, but the two most common tests are for:

1 IP2X – no contact may be made with a probe 12 mm in diameter and 80 mm long – in other words, a human finger

2 IP4X – no contact may be made with a rod of diameter 1 mm.

7.5.4 Tests for electrical separation of circuits [612.4.3]

This section is concerned with tests necessary to ensure the safety of safety extra-low voltage (SELV), protective extra-low voltage (PELV) and functional extra-low voltage (FELV) circuits which are explained in {8.17}. In general, the requirement is a thorough inspection to make sure that the source of low voltage (most usually a safety isolating transformer) complies in all respects with the British Standard concerned, followed by an insulation test between the extra-low voltage and low voltage systems. The test is unusual in that a 500 V dc supply (from an insulation resistance tester) must be applied between the systems for one minute, after which the insulation resistance must not be less than 5 MΩ for SELV or PELV systems, or 1.0 MΩ for FELV systems.

When insulation testing on electrically isolated circuits or on equipment which might be damaged by the test voltage, phase and neutral must be connected together and the test applied between them and earth.

7.6 EARTH TESTING

7.6.1 Testing earth electrodes [612.7]

The earth electrode, where used, is the means of making contact with the general mass of earth. Thus it must be tested to ensure that good contact is made. A major consideration here is to ensure that the electrode resistance is not so high that the voltage from earthed metalwork to earth exceeds 50 V. Where an RCD is used, this means that the result of multiplying the RCD operating current (in amperes) by the electrode resistance (in ohms) does not exceed 50 (volts) for normal dry locations. If a 30 mA RCD is used, this allows a maximum electrode resistance of 1,666 Ω, although it is recommended that earth electrode resistance should never be greater than 200 Ω. The minimum acceptable resistance where protection is by a 500 mA rated RCD is a value of 100 Ω. Whilst these values are acceptable, for reliability over a period of time, a minimum resistance of 200 Ω is suggested.

There are several methods for measurement of the earth electrode resistance. In all cases, the electrode must be disconnected from the earthing system of the installation before the tests commence.

1 Using a dedicated earth resistance tester

The instrument is connected as shown in {Fig 7.13} with terminals C_1 and P_1 being connected to the electrode under test (X). To ensure that the resistance of the test leads does not affect the result, separate leads should be used for these connections. If the test lead resistance is negligible, terminals C_1 and P_1 may be bridged at the instrument and connected to the earth electrode X with a single lead.

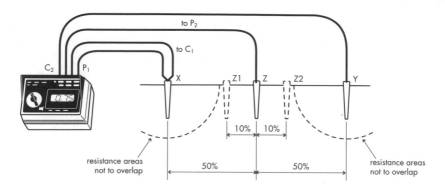

Fig 7.13 Measurement of earth electrode resistance with a dedicated tester

Terminals C_2 and P_2 are connected to temporary spikes Y and Z respectively that are driven into the ground, making a straight line with the electrode under test. It is important that the test spikes are far enough from each other and from the electrode under test. If their resistance areas overlap, the readings will differ for the reason indicated in {Fig 7.14}. Usually the distance from X to Y will be about 25 m, but this depends on the resistivity of the ground. To ensure that resistance areas do not overlap, second and third tests are made with the electrode Z 10% of the X to Y distance nearer to, and then 10% further from, X. If the three readings are substantially in agreement, this is the resistance of the electrode under test. If not, test electrodes Y and Z must be moved further from X and the tests repeated.

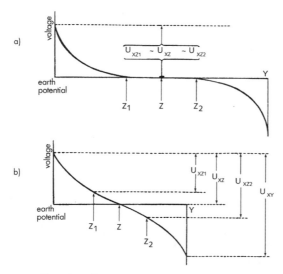

Fig 7.14 Effect of overlapping resistance areas
    a) resistance areas not overlapping
    b) resistance areas overlapping

The tester provides an alternating voltage output to prevent electrolytic effects. If the resistance to earth of the temporary spikes Y and Z is too high, a reduction is likely if they are driven deeper or if they are watered.

2 Using a transformer, ammeter and voltmeter

The system is connected as shown in {Fig 7.15}. Current, which can be adjusted by variation of the resistor R, is passed through the electrode under test (X) to the general mass of earth and hence to the test electrode Y. The voltmeter connected from X to Z measures the volt drop from X to the general mass of earth. The electrode resistance (Ω) is calculated from:

$$\frac{\text{voltmeter reading (V)}}{\text{ammeter reading (A)}}$$

As in the case of the dedicated tester, the test electrode Z must again be moved and extra readings taken to ensure that resistance areas do not overlap. It is important that the voltmeter used has high resistance (at least 200 Ω/V) since its low resistance in parallel with that of the electrode under test will give a false result.

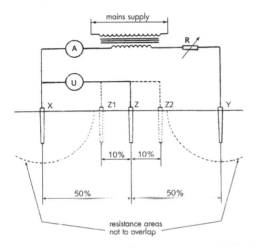

Fig 7.15 Measurement of earth electrode resistance with a transformer, ammeter and voltmeter

3 Using an earth-fault loop impedance tester

The tester is connected between the phase at the origin of the installation and the earth electrode under test as shown in {Fig 7.16}. The test is then carried out, the result being taken as the electrode resistance although the resistance of the protective system from the origin of the installation to the furthest point of the installation must be added to it before its use to verify that the 50 V level is not exceeded. If an RCD with a low operating current is used, the protective system resistance is likely to be negligible by comparison with the permissible electrode resistance.

It is most important to ensure that earthing leads and equipotential bonds are reconnected to the earth electrode when testing is complete.

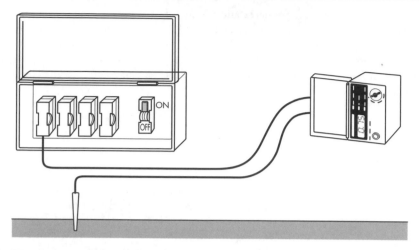

Fig 7.16 Measurement of earth electrode resistance with an earth-fault loop impedance tester

7.6.2 Measuring earth-fault loop impedance and prospective short-circuit current [612.9, Appendix 14]

The nature of the earth-fault loop and its significance have been considered in detail in {5.3}. Since the loop includes the resistance of phase and protective conductors within the installation, the highest values will occur at points furthest from the incoming supply position where these conductors are longest. A measurement within the installation will give the complete earth-fault loop impedance for the point at which it is taken (Z_S), or the earth-fault loop impedance external to the installation (Z_E) may be measured at the supply position. Internal loop measurements should be taken at points furthest from the intake to give the highest possible results.

In simple terms, the impedance of the phase-to-earth loop is measured by connecting a resistor (typically 10 Ω) from the phase to the protective conductor as shown in {Fig 7.17}. A fault current, usually something over 20 A, circulates in the fault loop, and the impedance of the loop is calculated within the instrument by dividing supply voltage by the value of this current. The resistance of the added resistor must be subtracted from this calculated value before the result is displayed. An alternative method is to measure the supply voltage both before and whilst the loop current is flowing. The difference is the volt drop in the loop due to the current, and loop impedance is calculated from voltage difference divided by current.

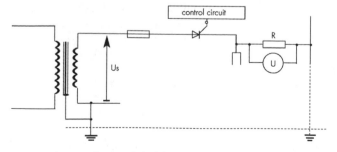

Fig 7.17 Simple principle of earth-fault loop testing

Since the loop current is very high, its duration must be short and must be limited to two cycles (or four half-cycles) or 40 ms for a 50 Hz supply. The current is usually switched by a thyristor or a triac, the firing time being controlled by an electronic timing circuit. It is very important to have already checked the continuity of the protective system before carrying out this test. A break in the protective system, or a high resistance within it, could otherwise result in the whole of the protective system being directly connected to the phase conductor for the duration of the test. Commercial testers are usually fitted with indicator lamps to confirm correct connection or to warn of reversed polarity. {Fig 7.18} shows a typical earth-fault loop tester connected to a socket outlet so that its loop impedance can be measured. Special leads for connection to phase and to earth are provided by suppliers for all other circuits.

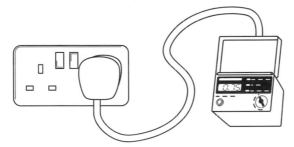

Fig 7.18 Earth-fault loop tester connected for use

Before testing, the main equipotential bonding conductors are disconnected (but not the connection with earth) to prevent parallel earth return paths and to ensure that there is no reliance on the service pipes for gas and water for effective earthing, (remember to reconnect the main equipotential bonding after the test).

Tests must be carried out at the origin of the installation, at each distribution board, at all fixed equipment, at all socket outlets, at 10% of all lighting outlets (choosing points furthest from the supply) and at the furthest point of every radial circuit. The test should be repeated at least once to allow for the effect of transient variations in the supply voltage.

A modified version of the earth-fault loop tester, which effectively measures the phase to neutral impedance and calculates then displays the value of the current which would flow if the supply voltage were applied to this impedance are readily available. The principle of such a PSC tester is described in {3.7.2}.

Since the test result is dependent on the supply voltage, small variations will affect the reading. Thus, the test should be repeated several times to ensure consistent results. The test resistor will be connected across the mains for the duration of each test, and will become very hot if frequent tests are made. Most testers will then 'lock out' to prevent further testing until the resistor temperature falls to a safe value.

The earth fault loop impedance measured as described will be for installation cables at ambient temperature, unless the circuit concerned has been in use immediately before the test, when it will be the impedance at normal operating temperature. Under normal operating conditions, cable temperature will rise, and so will the resistive component of the impedance. This effect is difficult to calculate, and a practical alternative is to ensure that the measured values of earth fault loop impedance do not exceed three quarters of the maximum values shown in {Tables 5.1, 5.2 or 5.4} as appropriate. A more complicated method of measuring earth fault loop impedance taking account of increased conductor resistances due to elevated temperatures is given in [Appendix 14].

The effect of supply voltage on the calculation of earth fault loop impedance is considered in {5.3.4}.

A circuit protected by an RCD will need special attention, because the earth-fault loop test will draw current from the phase which returns through the protective system. This will cause an RCD to trip. Therefore, any RCDs must be by-passed by short circuiting connections before earth-fault loop tests are carried out. It is, of course, of the greatest importance to ensure that such connections are removed after testing. Some manufacturers supply a loop tester which does not require RCDs to be short-circuited and which will not cause them to trip when the earth-fault loop test is made. Some instruments limit the test current to 15 mA so as not to trip RCDs with ratings of 30 mA and above. Whilst such tests may often be useful, they do not test the integrity of the system under fault current conditions.

When loop testing at lighting units controlled by passive infra-red detectors (PIRs), there may be damage to the associated electronic switches unless they are short-circuited before testing.

An alternative to the use of a dedicated earth-fault loop impedance tester is to measure the combined resistance of the phase and protective conductors from the incoming position to the point for which earth-fault loop impedance is required (this is $R_1 + R_2$ – see {7.4.4}) and to add to it the external earth-fault loop impedance (Z_E) which can be obtained from the electricity supplier. All earth-fault loop impedance test results should be carefully compared with the data in [Tables 41.2, 41.3 and 41.4], adjusted to allow for ambient temperature, or with figures provided by the designer. To ensure that ambient temperature is taken into account, the results should never exceed three quarters of the values given in the tables.

7.6.3 Testing residual current devices (RCDs) [612.10, Guidance Note 3]

Residual current devices should comply with BS EN 61008 and are described in {5.9}, from which it will be seen that they are provided with a built-in self-test system that is intended to be operated regularly by the user. BS 7671 requires that correct operation of this test facility should be checked, and that other tests are also carried out. The time taken for the device to operate must be measured, so the older type of 'go, no-go' tester is no longer adequate. {7.7.1} gives test instrument requirements.

RCD tests are carried out with a special tester that is connected between phase and protective conductors on the load side of the RCD after disconnecting the load {Fig 7.19}. A precisely measured current for a carefully timed period is drawn from the phase and returns via the earth, thus tripping the device. The tester measures and displays the exact time taken for the circuit to be opened. This time is very short, in most cases being between 10 and 20 ms, although it can be much longer, especially for S-types, which have delayed operation. For each of these tests. readings must be taken on both positive and negative half-cycles of the supply and the longer operating time recorded.

1 General purpose non-delayed RCDs

This is a general purpose type of RCD which is intended to operate very quickly at its rated current. Three tests are required:

   a) 50% of the rated tripping current applied for 2 s should not trip the device,
   b) 100% of rated tripping current, which should not be applied for more than
      2 s, must cause the device to trip within 200 ms (0.2 s), and

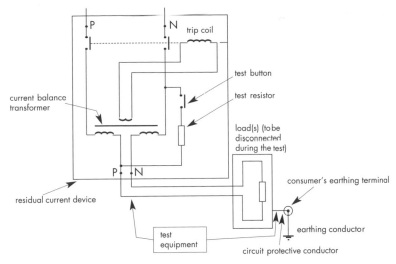

P   N trip coil

test button

current balance transformer

test resistor

load(s) (to be disconnected during the test)

P N

consumer's earthing terminal

residual current device

earthing conductor

test equipment

circuit protective conductor

Fig 7.19 Connections for an RCD tester

c) where the device is intended to provide supplementary protection against direct contact, a test current of 150 mA, applied for no more than 50 ms, should cause the device to operate within 40 ms.

2 Time delayed RCDs

In {5.9.2} we discussed the need for discrimination between RCDs. This type is deliberately delayed in its operation to make sure that other devices which are connected downstream of it will operate more quickly. A 3:1 discrimination ratio is required between two RCDs which are connected in series, and this must be verified before testing. It means that the delayed RCD must have an operating current at least three times that of the non-delayed type. For example, to discriminate properly with a 30 mA device, a second connected on the supply side would need to have an operating current of at least 90 mA (in practice, a 100 mA RCD is likely to be used).

The test for the time-delayed RCD consists of applying 100% of the normal rated current, when the device should trip within the time range of:

50% of rated time delay plus 200 ms, and
100% of rated time delay plus 200 ms.

For example, an RCD with a rated tripping time of 300 ms should trip within a time range of:

(150 + 200) ms = 350 ms
and (300 + 200) ms = 500 ms.

An RCD tester is an electronic device that draws current from the supply for its operation. This current is usually of the order of a few milliamperes, which is taken from the phase and neutral of the supply under test, and will have no effect on the measurement of single-phase systems. However, if a three-wire three-phase system (there is no neutral with this supply) is being tested, the tester must be connected to a neutral conductor to provide the power it needs for operation. Thus, its operating current will flow through a line conductor and return through the neutral, giving a basic imbalance. A 'no-trip' test must also be carried out, during which the RCD must not operate when 50% of the rated tripping current is applied for 2 s. The extra current to operate the tester, which adds to the test current, may then cause operation. It is necessary in this case to obtain from the RCD manufacturer the value of this current and to take it into account before failing a device on the 50% test.

The RCD tester is connected to the device to be tested by plugging it into a suitable socket outlet (see {Fig 7.20}) or by connecting to phase and neutral with special leads obtainable from the instrument supplier.

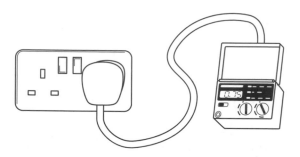

Fig 7.20 RCD tester connected for use,

7.7 TEST INSTRUMENT REQUIREMENTS

7.7.1 Basic requirements [612.1]

Guidance Note 3 – Inspection and Testing – makes it clear that instrument accuracy is required to be at least that shown later in this sub-section for the various types of instrument. Regular recalibration using standards traceable to National Standards is now required, together with checking after any incident which has involved mechanical mishandling. Many electrical installers will not be used to sending their instruments regularly for recalibration, but must now do so. At least one instrument manufacturer supplies a box (a check box) containing a series of accurate resistors for use in checking the performance of instruments. Whilst not taking the place of a properly verified recalibration, such a device can prove useful between calibrations to make sure that instruments are not wildly inaccurate. Guidance Note 3 is not specific on the time intervals at which recalibration must be carried out, but it would seem sensible for occasionally used instruments to receive attention every year, whilst those used frequently are likely to need recalibration more frequently. Instruments should also be recalibrated following damage due to dropping or due to excess current.

If installations are to be tested to show that they comply with BS 7671, the following instruments will be necessary. After the name of the instrument are brief notes which may be helpful in choosing a new instrument or in deciding if one already to hand will be satisfactory. The first four instruments listed are absolutely essential for all tests, although the low resistance tester and the insulation resistance tester may be combined in a single instrument. An instrument which can test continuity, insulation resistance, earth-fault loop impedance and the operation of RCDs is available from many manufacturers, and will save a great deal of time in changing instruments and test leads – even more time can be saved with an instrument which will save the test results for down-loading to a computer. The last two instruments will not often be required on simple installations, since applied voltage and earth electrode resistance tests are seldom needed. Test leads, including prods and clips, must be in good order and have no cracked or broken insulation. Fused test leads are recommended to reduce the risk of arcing under fault conditions.

The basic accuracy of ±5% quoted below is for digital instruments. Analogue instruments should have an accuracy of ±2% at full scale deflection, which will give the required accuracy of ±5% over the useful part of their scales.

Low resistance ohmmeter
Basic instrument accuracy required is ±5%
Test voltage ac or dc, between 4 V and 24 V
    Test (short circuit) current not less than 200 mA
    Able to measure from 0.2 Ω to 2 Ω to within 0.01 Ω (resolution of 0.01)
    Facility to store and subtract test leads resistance will be valuable
    May be the continuity range of an insulation resistance tester
    Instruments to BS EN 61557-4 will comply

Insulation resistance tester
    Direct test voltage depends on the circuit under test, but will be:-
        250 V for extra-low voltage circuits
        500 V for other circuits supplied at up to 500 V
        1000 V for circuits rated between 500 V and 1000 V.
    Must be capable of delivering a current of 1 mA at the minimum allowable resistance level, which is:
        0.5 MΩ for the 250 V tester
        1.0 MΩ for the 500 V tester
        1.0 MΩ for the 1000 V tester
    Basic instrument accuracy required is ±5%
    Must have a facility to discharge capacitance up to 5 μF which has become charged during the test
    May be combined with the low resistance ohmmeter
    Instruments to BS EN 61557-2 will comply

Earth-fault loop impedance tester
    Must provide 20 to 25 A for up to two cycles or four half-cycles
    Basic instrument accuracy required is ±5%
    Able to measure to within 0.01 Ω (resolution of 0.01)
    Instruments to BS EN 61557-3 will comply

Residual current device (RCD) tester
    Must perform the required range of tests (see {7.6.3})
    Suitable for standard RCD ratings of 6, 10, 30, 100, 300 and 500 mA
    Must NOT apply full rated test current for more than 2 s
    Currents applied must be accurate to within ±10%
    Able to measure time to within 1 ms (resolution of 1)
    Must measure opening time with an accuracy of ±5%
    Instruments to BS EN 61557-6 will comply

Applied voltage tester
    Must apply a steadily increasing voltage measured with an accuracy of ±5%
    Must have means of indicating when insulation breakdown has occurred
    Must be able to maintain the test voltage for at least one minute
    Maximum output current must not exceed 5 mA
    Maximum output voltage required is 4000 V

Earth electrode resistance tester
    Basic instrument accuracy required is ±5%
    Must include facility to check that the resistance to earth of temporary test spikes are within limits
    Able to measure to within 0.01 Ω (resolution of 0.01)

7.7.2 Accuracy and resolution

The sub-section above has indicated the levels of accuracy and resolution required of the instruments needed to test an electrical installation. The purpose of this sub-section will be to explain the meaning of these two terms.

Accuracy

This term describes how closely the instrument is able to produce and display a correct result, and is usually expressed as a percentage. For example, if a voltage has a true level of 100 V and is measured by a voltmeter as 97 V, this is an error of –3 V. Expressed in terms of the true voltage, it is an error of three volts in one hundred volts, or three per cent (3%). In this case the reading is low, so the true error would be –3%. Had the error been +3% the reading would have been 103 V. In most cases we do not know if the reading is high or low, so the error is expressed as a percentage that may be positive or negative. Thus, if the voltmeter gave a reading of 100 V but was known to have an accuracy of ±4%, the actual voltage could lie anywhere in the range from:

$$100 + (4/100) \times 100\,V \quad \text{to} \quad 100 - (4/100) \times 100\,V$$
$$\text{or} \quad 100 + 4\,V \quad\quad\quad \text{to} \quad 100 - 4\,V$$

which is between 104 V and 96 V.

The values given in the Guide are called basic instrument accuracies that indicate the possible error with the instrument itself. In practice, there are many factors affecting the value which is to be measured, and which will further reduce the accuracy. These are divided into two types.

Instrument errors are largely due to the fact that the true error is not constant, varying from point to point over the instrument range. Other factors, such as battery voltage, ambient temperature, operator's competence, and the position in which the instrument is held or placed (such as vertical or horizontal) will also affect the reading.

Field errors concern external influences which may also reduce accuracy, and may include capacitance in the test object, external magnetic fields due to cables and equipment, variations in mains voltage during the test period, test lead resistance, contact resistance, mains pickup, thermocouple effects, and so on.

It is important to appreciate that percentage accuracy is taken in terms of the full scale reading of an analogue instrument, or the highest possible reading of a digital type. Thus, in a multirange instrument, it is related to the scale employed, not to the reading taken. For example, if an ohmmeter with a known error of ±5% is on its 100 Ω scale and reads 8 Ω, the true reading will lie between

$$8 + \frac{100 \times 5}{100} = 8 + 5\,\Omega = 13\,\Omega \text{ and}$$
$$8 - \frac{100 \times 5}{100} = 8 - 5\,\Omega = 3\,\Omega$$

and **not** between $8 \pm \dfrac{8 \times 5}{100} = 8 \pm 0.4\,\Omega$ or 7.6 Ω and 8.4 Ω.

Thus it can be seen that the highest accuracy will result from using the lowest possible scale on a multirange instrument.

Resolution

This term deals with the ability of an instrument to display a reading to the required degree of accuracy. For example, consider checking the earth-fault loop impedance of a circuit expected to be 1.14 Ω. If this were done using a digital meter with three digits and a lowest range of 99.9 Ω, we could obtain a reading of 1.1 Ω or another of 1.2 Ω, but not 1.14 Ω. This would indicate that the instrument resolution was to the nearest 0.1 Ω, which usually is not close enough for electrical installation measurements. If the same three-digit instrument had a lower scale of 9.99 Ω, it would be capable of reading 1.14 Ω and would have a resolution of 0.01.

7.8 SUPPORTING PAPERWORK

7.8.1 Why bother with paperwork? [631, Appendix 6]

When an installation is complete, including additions or alterations to an existing installation, the person responsible for the work must report to the owner that it is complete and ready for service. This is in the form of an electrical installation certificate complete with a record of inspection and the results of testing, which must be separately signed to verify the design, the construction and the inspection and test aspects (see {7.8.2}) to confirm that the installation complies with BS 7671 as well as with the designer's intentions. Sometimes, the person ordering the work is not the user, in which case it must be ensured that the user has copies of the forms for inclusion in the installation manual.

Domestic installations are now subject to the Building Regulations and special rules apply as far as notifying the Local Authority and providing test certificates are concerned. The requirements are explained in {2.2.2} and in greater detail in {Chapter11}.

This certificate will verify that the installation is safe to use and ready for service, and should be signed by a competent person who should preferably be one of the following:

1 a professionally qualified electrical engineer, or
2 a member of the ECA (Electrical Contractors' Association), or
3 a member of the ECA of Scotland, or
4 an approved contractor of the NICEIC (National Inspection Council for Electrical Installation Contracting), or
5 a qualified person acting on behalf of one of the above.

In the case of the first testing of a new installation, it is important that the results are compared with the design criteria to ensure that the installation has been completed as intended. If the installation is being re-tested, the results must be compared with those of the previous tests so that deterioration can be pinpointed.

In all cases, the Certificates must state for whom the qualified person is acting. The installer should also compile an operational manual for the installation, which will include all the relevant data, including:

1 a full set of circuit and schematic drawings,
2 all design calculations for cable sizes, cable volt drop, earth-loop impedance, etc.
3 leaflets or manufacturers' details for all the equipment installed,
4 'as fitted' drawings of the completed work where applicable,
5 a full specification,
6 copies of the electrical installation certificate, together with any other commissioning records,

7   a schedule of dates for periodic inspection and testing,
8   the names, addresses and telephone numbers of the designer, the installer, and
    the inspector/tester.

This requirement will be comparatively new to many electrical installation installers. Its rationale is to ensure that future owners or users of the installation, as well as those who maintain it or who may modify it, have full information. If difficulty is experienced in preparing the operational manual, reference to BS 4884 and to BS 4940, may be helpful. Records of the results of all tests (in the form of Electrical Installation and Periodic Test and Inspection Certificates) should be kept, ideally with the Operational Manual, for the life of the installation. For the first time the 2004 Amendments allow the test and reporting data to be kept as an electronic (i.e. computer) record, provided that they are durable. Presumably this means that they must be kept on a compact disk (CD) or other secure storage device, and not simply in the volatile computer memory. Before long it seems that anyone selling a domestic property will be required to supply full details of its design, construction and history in the form of a home information pack.

7.8.2 Electrical installation certificate [631, Appendix 6]

Following completion of inspection and testing, an electrical installation certificate, together with particulars of the electrical installation and a schedule of test results, must be provided to the person who ordered the work. It must be signed three times (by the designer, the installer or constructor and the inspector/tester) to certify that the installation has been designed, constructed, inspected and tested in accordance with BS 7671:2008 (the 17th Edition of the IEE Wiring Regulations) but a single-signature version of the certificate may be used where design, construction, inspection and testing are all the work of the same person. The form of the certificate giving particulars of the electrical installation and of completion and inspection is given as {Table 7.9}. In some cases, where there are good reasons and where a qualified electrical engineer has given his approval, the installation may not comply fully with the Regulations. In such a case, full details of the departures must be stated on the completion and inspection certificate as indicated in {Table 7.9}, although this table may be expanded and redesigned as necessary to cover the particular installation under test. A full schedule of test results and inspections must be appended to the Electrical Installation Certificate and the Periodic Inspection Report as shown in {Table 7.10}. The complete certificate must be handed (or sent) to the person ordering the work, together with a copy to be given to the user of the installation where different (see {Table 7.9} for the details of the Electrical Installation Certificate). The installer (constructor), designer and tester should also keep copies for their records. The recommended intervals between periodic inspections and tests should be as indicated in {Table 7.4}. It seems almost certain that the advent of the ESQC Regulations (see {2.2.2}) will mean that the installer must provide an electrical installation certificate to the Electricity Supplier before he will connect the installation to his system.

7.8.3 Installation alterations and additions [633]

The changes in the occupation and uses of buildings where there are electrical installations make alterations and additions a common occurrence. Before commencing alterations and additions to an installation it is very important to verify that the existing supply system, installation and earthing arrangements are capable of feeding the proposed new installation safely. For example, the additional installation in a large extension to a house may impose extra loads on the supply system that it is incapable of meeting. It is the responsibility of the person carrying out the extra work to ensure that it, as well as the existing installation, will function safely and correctly. The tester must check that all materials made redundant by the installation changes, such as cables, wooden pattresses, etc., and which may be responsible for the spread of fire, are removed. Additionally, the inspector must look for other prospective fire risks, such as:

1 flexible cords not securely held by cord grips,
2 a dangerous increase in the use of adaptors,
3 worn or otherwise damaged flexible cords,
4 signs of overheating of appliances, plugs and connectors,
5 over-rated lamps fitted to luminaires, and
6 heaters with insecure or missing guards.

It is not his responsibility to rectify defects and faults in the existing installation, but he must test and inspect it, reporting deviations on the electrical installation certificate that is provided on completion of the work. Should dangerous defects be found in an existing installation it would be clearly irresponsible of the person carrying out the alterations or extensions not to bring them to the urgent attention of the user. On completion of alterations or additions to an electrical installation an Electrical Installation Certificate (see{Table 7.9}) must be issued.

7.8.4 Periodic inspection and testing [62]

The importance of regular inspection and testing of electrical installations cannot be overstated, but unfortunately it is an aspect of electrical safety that is very often overlooked. It is now a requirement of the Regulations that the installer of an installation must tell the user of the need for periodic test and inspection and the date on which such attention is required. Probably the good contractor will institute a system so that a reminder is sent to the customer at the right time. Suggested intervals between inspections and tests are shown in {Table 7.4}. The results of sample tests should be compared with those taken when the installation was last tested and any differences noted. Unless the reasons for such differences can be clearly identified as relating only to the sample concerned, more tests must be carried out. If these, too, fail to comply with the required values, the complete installation must be retested and the necessary correcting action taken.

The periodic inspection and testing must be carried out with the same degree of care as is required for a new installation. In fact, more care is often needed because dangers can occur to the testers and to others in the situation in the event of failure of parts of an installation such as the protective system. The tester must look out for additions to the installation, or for changes in the use of the area it serves, either of which may give rise to fire risks. Included may be the addition of thermal insulation, the installation of additional cables in conduit or trunking, dust or dirt which restricts ventilation openings or forms an explosive mixture with air, changing lamps for others of higher rating, missing covers on joint boxes and other enclosures so that vermin may attack cables, and so on.

The sequence of periodic tests differs slightly from that for initial tests {7.3.2} because in this case the supply will always be connected before testing starts. The tests required and the sequence in which they must be performed are shown in {Table 7.13}. There will be cases where necessary information in the form of charts,

Table 7.9 ELECTRICAL INSTALLATION CERTIFICATE (BS 7671: 2008)
(From Appendix 6 of BS 7671: 2008)

DETAILS OF THE INSTALLATION
Client's name/title ....
Installation address ....
Extent of installation covered by this certificate ....
(Use continuation sheet if necessary)

DESIGN
I/we being the person(s) responsible (as indicated by my/our signatures below) for the design of the electrical installation, particulars of which are described on page 1, have exercised reasonable skill and care in carrying out the design hereby CERTIFY that the said work for which I/we have been responsible is to the best of my/our knowledge and belief in accordance with BS 7671: 2008 – Requirements for Electrical Installations (17th Edition IEE Wiring Regulations) amended to .... except for the departures, if any, stated in this certificate.
Details of departures (if any) from BS 7671: 2008
For the DESIGN of the installation
Name (in BLOCK letters) .... Position ....
Signature ....
Date ....
For and on behalf of ....
Address: ....

CONSTRUCTION
I/we being the person(s) responsible (as indicated by my/our signatures below) for the construction of the electrical installation, particulars of which are described on page 1, have exercised reasonable skill and care in carrying out the construction hereby CERTIFY that the said work for which I/we have been responsible is to the best of my/our knowledge and belief in accordance with BS 7671: 2008 – Requirements for Electrical Installations (17th Edition IEE Wiring Regulations) amended to .... except for the departures, if any, stated in this certificate.
Details of departures from BS 7671: 2008 ....
....
The extent of liability (if any) of the signatory is limited to the work described above as the subject of this Certificate
For the CONSTRUCTION of the installation
Name (in BLOCK letters) .... Position ....
Signature .... Date ....
For and on behalf of ....
Address ....

INSPECTION AND TESTING
I/we being the person(s) responsible (as indicated by my/our signatures below) for the inspection and test of the electrical installation, particulars of which are described above, have exercised reasonable skill and care in carrying out the inspection and testing hereby CERTIFY that the said work for which I/we have been responsible is to the best of my/our knowledge and belief in accordance with BS 7671: 2008 – Requirements for Electrical Installations (17th Edition IEE Wiring Regulations) amended to .... except for the departures, if any, stated in this certificate.
The extent of liability of the signatory is limited to the work described as the subject of this Certificate.
For the INSPECTION AND TESTING of the installation
Name (in BLOCK letters) .... Position ....
Signature .... Date ....
For and on behalf of ....
Address ....
I/we the designer(s) RECOMMEND that this installation be further inspected and tested after an interval of not more than .... months/years

PARTICULARS OF THE INSTALLATION

Type of earthing TN-C ☐   TN-S ☐   TN-C-S ☐   TT ☐   IT ☐

Distributors facility ☐   Installation earth electrode ☐

Number and type of live conductors ....

Nature of supply parameters

Nominal voltage .... Frequency .... Hz

Prospective fault current I_{pf} .... kA.   External loop impedance Z_e .... Ω

          Measured ☐   Calculated ☐   Other ☐

Supply protective device characteristics   Type .... Nominal current rating .... A

Earthing – Distributors facility .... Earth electrode ....

Maximum demand (load) .... A

Earth electrode details (where applicable)

Type .... Location .... Resistance to earth .... Ω

Main protective conductors:

Earthing conductor material .... csa .... connection verified

Main equipotential

Bonding conductors: material .... csa .... connection verified

To incoming water and/or gas service .... To other elements ....

Main switch or circuit breaker:

Number of poles .... Type BS .... Rating .... A Voltage rating V

Location .... Fuse rating or setting .... A

(if a residual current device, rated residual operating current .... mA, operating time

.... ms

Comments on existing installation, in case of an alteration or addition

....

Schedule of Circuits and Test Results: this Certificate is valid only if schedules are attached.

Schedules of Inspections and Schedules of Test Results are attached (see {Table 7.10}).

diagrams, tables, etc., is not available. The tester will then need to investigate the installation more thoroughly to make sure that he is fully conversant with it before he can carry out his work.

It must be clearly understood that a retest of a working installation may not be as full as that carried out before the installation was put into service. For example, it may be impossible to switch off certain systems such as computers, and if full testing is not possible, this should be made clear on the certificate. In such cases, the installation should be carefully inspected to ensure that possible dangers become apparent. In some cases it may be necessary to carry out measurements, such as the value of the earth leakage current, which will indicate the health of the system without the need to disconnect it.

A sample of 10% of switching devices must be thoroughly internally inspected and tested. If results are poor, the procedure must be extended to include all switches. The condition of conductor insulation and other protection against direct contact must also be inspected at all distribution boards and at samples of switchgear, luminaires, socket outlets, etc. There should be no signs of damage, overloading or overheating. Earth-fault loop impedance must be tested at the origin of the installation, at each distribution board, at 10% of socket outlets on ring circuits and at the extremity of every radial circuit.

Protective and equipotential bonding conductors must not be disconnected from the main earthing terminal unless it is possible first to isolate the supply. The tester must ensure that a durable notice to the wording given in {Table 7.12} is fixed at the mains position.

On completion a full electrical installation certificate must be submitted. Additional matters to report are:

1 full test results to enable comparison with earlier tests, from which the rate of deterioration of the installation (if any) can be assessed. The required form form, is given as {Table 7.11}

2 the full extent of the parts of the system tested - notes of omissions may be very important

3 any restrictions that may have been imposed on the tester and which may have limited his ability to report fully

4 any dangerous conditions found during testing and inspection, non-compliance with the Regulations, or any variations that are likely to arise in the future.

Table 7.10 Schedule of Test Results

Contractor ....
Test Date ....
Signature ....
Method of fault protection ....
Equipment vulnerable to testing ....
Description of Work ....

Address/location of distribution board ....
Type of supply: TN-S/TN-C-S/TT
Z_e at origin Ω
PFC (kA)
Confirmation of supply polarity □

Instruments
loop impedance
continuity
insulation
RCD tester

Circuit Description	Overcurrent Device Short-circuit capacity: ....kA		Wiring Conductors		Continuity		Insulation Resistance		Polarity	Earth Loop Impedance	Functional Testing		Remarks
	type	rating I_n A	live mm²	cpc mm²	(R_1+R_2)* Ω	R_2* Ω	Live/ Live MΩ	Live/ Earth MΩ		Z_s Ω	RCD time ms	RCD Other	
1	2	3	4	5	6	7	9	10	11	12	13	14	15

Deviations from BS 7671: IEE Wiring Regulations and special notes:

Table 7.11 PERIODIC INSPECTION REPORT FOR AN ELECTRICAL INSTALLATION (Appendix 6 of BS 7671: 2008)

DETAILS OF THE CLIENT
Client ....
Address ....
Purpose for which the report is required
....

DETAILS OF THE INSTALLATION
Occupier .... Address ....
....
....
Description of premises:　☐ Domestic　☐ Commercial　☐ Industrial
Other ....
Estimated age of the electrical installation .... years
Evidence of alterations or additions　☐ Yes　☐ No　☐ Not apparent
If 'yes', estimated age .... years.
Date of last inspection .... Records available　☐ Yes　☐ No
Records held by ....

EXTENT AND LIMITATIONS OF THE INSPECTION
Extent of electrical installation covered by this report:
....
....
Limitations: ....
....

This inspection has been carried out in accordance with BS 7671: 2008 (IEE Wiring Regulations), amended to .... Cables concealed within trunking and conduits, or cables and conduits concealed under floors, in roof spaces and generally within the fabric of the building or underground have not been inspected.

NEXT INSPECTION
We recommend that the installation should be re-inspected and tested after an interval of not more than .... months/years providing the matters requiring urgent attention are attended to without delay.

DECLARATION – INSPECTED AND TESTED BY
Name .... Signature ....
For and on behalf of .... Position ....
Address ....
.... Date ....

SUPPLY CHARACTERISTICS AND EARTHING ARRANGEMENTS
Type of earthing　TN-C ☐　　TN-S ☐　　TN-C-S ☐　　TT ☐　　IT ☐
Number and type of live conductors ....
Nominal voltage .... Frequency .... Hz
Prospective fault current I_{pf} .... kA　External loop impedance Z_e .... Ω
　　Measured ☐　　Calculated ☐　　Other ☐
Supply protective device characteristics Type .... Nominal current rating .... A
Maximum demand (load) .... A
Means of earthing Distributors facility　Earth electrode
Earth electrode details (where applicable)
Type .... Location .... Resistance to earth .... Ω
Main protective conductors:
Earthing conductor material .... csa ....
Main equipotential bonding conductors: material .... csa ....
To incoming water and/or gas service
To other elements ....

Main switch or circuit breaker:
Number of poles .... Type BS .... Rating .... A Voltage rating V
Location .... Fuse rating or setting .... A
(if a residual current device, rated residual operating current .... mA, operating time
.... ms

OBSERVATIONS AND RECOMMENDATIONS
Referring to the attached Schedule(s) of Inspection and Test Results, and subject to the limitations specified,
No remedial work is required or the following observations are made
....
....
One of the following numbers shall be placed alongside each of the items detailed above:
1. requires urgent attention   2. requires improvements   3. requires further investigation
    4. does not comply with the current BS 7671 (as amended). (This does not necessarily
    imply that the electrical installation is unsafe).

Comments on existing installation, in case of an alteration or addition
....

SUMMARY OF THE INSPECTION
Date(s) of inspection ....
General condition of the installation ....
....
Overall assessment: Satisfactory/Unsatisfactory

DECLARATION
INSPECTED AND TESTED BY:
Name .... Signature .... Date ....
For and on behalf of .... Position...
Address ....
The Schedule of Circuits and Test Results: this Certificate is valid only if schedules are attached.
.... Schedules of Inspections and .... Schedules of Test Results are attached (see
{Table 7.10})

Where an installation was constructed to comply with an earlier Edition of the
Regulations, tests should be made as required by the 17th Edition as far as it is
applicable and the position fully explained, with suggestions of necessary action, in
the Electrical Installation report.

Table 7.12 Notice – periodic inspection and testing

IMPORTANT
This installation should be periodically inspected and tested and a report on its condition
obtained, as prescribed in BS 7671 (formerly The IEE Wiring Regulations for Electrical
Installations) published by the Institution of Electrical Engineers.

Date of last inspection ....

Recommended date of next inspection ....

Table 7.13 Schedule of inspections (From Appendix 6 of BS 7671: 2008)

Methods of protection against electric shock
a) Basic and fault protection
   i) SELV
   ii) Limitation of discharge of energy
b) Basic protection
   i) Insulation of live parts
   ii) Barriers or enclosures
   ii) Obstacles
   iv) Placing out of reach
   v) PELV
   vi) Presence of RCD for additional protection
c) Fault protection
   i) EEBADS including
      Presence of earthing conductor
      Presence of circuit protective conductors
      Presence of main equipotential bonding conductors
      Presence of supplementary equipotential bonding conductors
      Presence of earthing arrangements for combined protective and functional purposes
      Presence of adequate arrangements for alternative source(s) where applicable
      Presence of residual current device(s)
   ii) Use of Class II equipment or equivalent insulation
   iii) Non-conducting location: absence of protective conductors
   iv) Earth-free equipotential bonding: presence of suitable conductors
   v) Electrical separation

Prevention of mutual detrimental influence
a) Proximity of non-electrical services and other influences
b) Segregation of Band I and Band II circuits or Band II insulation used
c) Segregation of safety circuits

Identification
a) Presence of diagrams, instructions, circuit charts and similar information
b) Presence of danger notices and other warning notice
c) Labelling of protective devices, switches and terminals
d) Identification of conductors

Cables and conductors
a) Routing of cables in prescribed zones or within mechanical protection
b) Connection of conductors
c) Erection methods
d) Selection of conductors for current carrying capacity and volt drop
e) Presence of fire barriers, suitable seals and protection against thermal effects

General
a) Presence and correct location of appropriate devices for isolation and switching
b) Adequacy of access to switchgear and other equipment
c) Particular protective measures for special installations and locations
d) Connection of single-pole devices for protection or switching in line conductors only
e) Correct connection of accessories and equipment
f) Presence of undervoltage protection devices
g) Choice and setting of protective and monitoring devices for basic protection and/or overcurrent protection
h) Selection of equipment and protective measures appropriate to external influences
i) Selection of appropriate functional switching devices

Inspected by .... Date ....

7.8.5 Minor electrical installation works certificate [631, Appendix 6]

This simplified certificate is provided by the installer after completion of additions, alterations or replacements to the existing electrical installation which do not extend to the provision of a new circuit. Examples include the addition of a socket outlet or a lighting point to an existing circuit, or to the replacement or repositioning of a light switch. A separate certificate must be provided for each circuit on which work has been carried out. The details required on this certificate are shown in {Table 7.14}.

Table 7.14 Minor electrical installation works certificate
(from [Appendix 6] of BS 7671: 2008)

Part 1 DESCRIPTION OF MINOR WORKS
Description of minor works ....
Location/Address ....
Date minor works completed ....
Details of departures, if any, from BS 7671 ....

Part 2 INSTALLATION DETAILS
1 System earthing arrangement TN-C-S ☐   TN-S ☐   TT ☐
Method of fault protection ....
Protective devices for the modified circuit Type .... Rating .... A
Comments on existing installation, including adequacy of earthing and bonding [131.8]
....
....

Part 3 ESSENTIAL TESTS
        Earth continuity satisfactory ....
        Insulation resistance   Phase/neutral .... MΩ
                                Phase/earth   .... MΩ
                                Neutral/earth .... MΩ
        Earth fault loop impedance .... Ω
        Polarity satisfactory ....
        RCD operation. Rated residual operating current .... mA operating time
        .... ms.

Part 4 DECLARATION
We/I CERTIFY that the said works do not impair the safety of the existing installation, that the said works have been designed, constructed, inspected and tested in accordance with BS 7671: 2008 (IEE Wiring Regulations), and that, to the best of my/our knowledge and belief, at the time of my/our inspection, complied with BS 7671: 2008 except as detailed in Part 1.
        Signature: .... Position: ....
        Name: .... Date: ....
        Company ....
        Address ....
        ....

NOTES FOR RECIPIENTS The Minor Works Form is only to be used for additions or alterations to an installation that does not extend to the provision of a new circuit.

CHAPTER EIGHT

Special installations

8.1 INTRODUCTION [700]

The Regulations apply to all electrical installations in buildings. There are some situations out-of-doors, as well as indoor ones, which are the subject of special requirements due to the extra dangers they pose, and these will be considered in this chapter. These Regulations are additional to all of the other requirements, and not alternatives to them. [Part 8] has some strange non-sequential Regulation numbers because of the need to align with international work.

8.2 BATH TUBS AND SHOWER BASINS [701]

8.2.1 Introduction [701.1]

People using a bathroom are often unclothed and wet. The absence of clothing (particularly shoes) will remove much of their protection from shock (see {3.4}), whilst the water on their skin will tend to short-circuit its natural protection. Thus, such people are very vulnerable to electric shock due to their reduced body resistance, so special measures are needed to ensure that the possibility of basic or fault contact (previously described as direct or indirect contact) is much reduced. Only separated extra-low voltage (SELV) or protected extra-low voltage (PELV) may be installed, other than water heaters in zones 1 and 2, and equipment specially designed to be safe in the vicinity of swimming pools. Attention is drawn to the fact that BS 7671:2008 does NOT apply to natural areas such as lakes, gravel pits and so on. Medical pools are also outside the scope of this Section, but see {8.19}. Guidance Note 7 (Special Locations) provides data on the impedance of the human body. However, the figures are complicated by the fact that values differ significantly from person to person and with applied voltage; it would be sensible to assume a worst-case possibility which suggests that the impedance of the human body from hand to foot is as low as 500 Ω.

Since this calculates to a body current of 460 mA at 230 V we are considering a fatal shock situation. When the fact that the impedance values assume a dry body are taken into account and may be halved for wet people, we come to a shock current approaching 1 A! Attention is drawn to the fact that a bath used for medical treatment or for disabled people may need special consideration, because the possible hazard is greater.

The special requirements of this section do not apply to rooms (such as bedrooms) containing an enclosed and prefabricated shower basin, provided that switches are not mounted within 0.6 m or sockets within 2.4 m of the shower door opening. All socket outlets in such rooms must be protected by an RCD with a rating no higher than 30 mA. The special requirements do not apply to emergency facilities in industrial areas and laboratories or to areas for medical treatment or use by disabled persons. These are covered in {8.19.2}, Medical Locations.

8.2.2 Bath and shower room requirements [701]

The revised Regulations for bath and shower rooms are based on the dimensions of three zones into which the space is divided. This marks a simplification, since the 16th Edition specified four zones.

> **Zone 0** is the interior of the bathtub or the shower basin. In those cases where a shower has no basin, the zone extends 100 mm above the floor and includes the volume within the vertical surface at a radius of 1.2 m from the water outlet position of a fixed shower-head; in fact, the radius of 1.2 m includes both zone 0 and zone 1 (see {Fig 8.1a to 8.1e}). There is no mention of the very common demountable showerhead.
>
> **Zone 1** extends from Zone 0 up to 2.25 m above the floor, and outwards to the extent of the bath or basin, or the extent of Zone 0 as defined above where there is no shower basin. It includes the space below a bath or shower tub or basin where that space is accessible without the use of a tool; if a tool must be used to gain access to such a space, it is outside all zones.
>
> **Zone 2** extends outside Zone 1 for 0.6 m horizontally and up to a height of 2.25 m. When a window is in both zones 1 and 2, its recess is counted as being in zone 2. In the common case of a permanently fixed partition adjacent to the bath or the shower, the extent of zone 2 is shown in {Fig 8.1c}.
>
> The extent of the zones is shown in {Figs 8.1a to 8.1e} in diagrams that are not to scale.

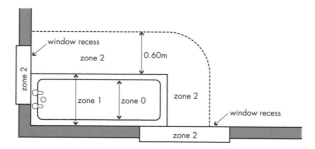

Fig 8.1a Zones for a bathroom (plan). Note that the same zones apply to a shower room with a shower basin.

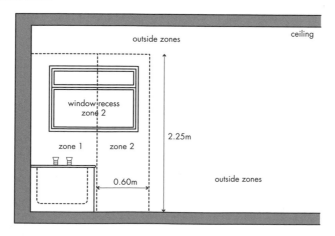

Fig 8.1b Zones for a bathroom (elevation). Note that the same zones apply to a shower room with a shower basin

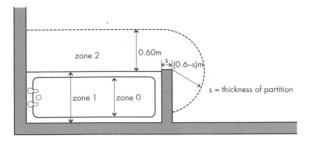

Fig 8.1c Zones for a bathroom (plan) with a permanently fixed partition. Note that the same zones apply to a shower room with a shower basin.

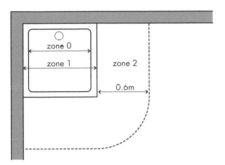

Fig 8.1d Zones for a shower (plan) with a basin and a fixed showerhead

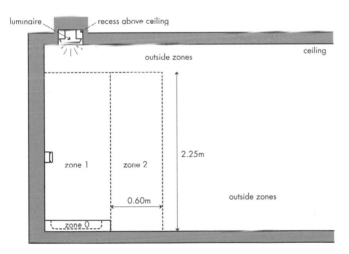

Fig 8.1e Zones for a shower (elevation) with a basin and fixed showerhead

Where SELV (separated extra-low voltage) or PELV (protective extra-low voltage), see ({3.4.4}) is used, protection against basic shock, it must be provided by barriers or enclosures to IP2X (no finger contact) or by insulation that will withstand 500 V a.c. for one minute. Local supplementary equipotential bonding must be installed to connect together the exposed conductive parts of Class I and Class II equipment in Zones 1 and 2, together with extraneous conductive parts, such as metal baths and shower trays, metallic pipes for water, gas, central heating, air conditioning, etc.

in all three zones but not outside them. Accessible structural metal must also be bonded, although this is unnecessary in the cases of metal door architraves, window frames and so on, unless these parts are connected to the metal structure of the building, which in practice is unlikely. These bonding requirements do not apply where a prefabricated shower or bath is installed in a bedroom.

The use of plastic piping for water and central heating services is now very common. Where such piping is used in bathrooms, the requirements for supplementary bonding are relaxed. No bonding is required for taps, pipes or central heating radiators, but must still be used for electrical equipment such as shaver units, electric showers, luminaires, etc.

In some cases, separated extra-low voltage (SELV) at 12 V a.c. or 30 V ripple-free d.c. is used within the bath and shower basin (for example for hoists used by disabled persons). Such systems must be protected against water ingression by being rated IPX7, suitable for immersion. The safety source supplying such a SELV system must be situated outside the zones. Protections by obstacles, placing out of reach {3.4.5}, or a non-conducting location {5.8.2} are not allowed in bath or shower rooms.

Switch and control gear is permitted in any zone if such gear is incorporated in the equipment installed there, which is suitable for use in that zone. No other switch or control gear is permitted in Zone 0. In Zone 1, only switch and control gear for SELV may be installed, but the safety source for the system must be installed outside the zones. In Zone 2, or outside the zones but within the bath or shower room, switches and socket outlets are permitted provided that they are fed from an SELV source installed outside the zones. The exception to this is the socket outlet on a shaver supply unit that complies with BS EN 61778-2-5, which may be installed in Zone 2. The 17th Edition allows the installation of socket outlets in bathrooms for the first time, provided that they are 3 m horizontally beyond the boundary of Zone 1. It will be appreciated that the bathroom that allowed such sockets to be installed would be unusually large. Insulated operating cords of pull switches are permitted in Zones 1 and 2, but they must comply with BS EN 60669-1. All equipment installed in all zones must be protected by one or more RCDs with ratings not exceeding 30 mA.

Electrical equipment installed in a room containing a bath or shower must have protection against the ingress of water as follows:

In Zone 0, only equipment which is intended for that situation, water protected to IPX7 or

In Zones 1 and 2, IPX4 which is protection from splashing. Where water jets may be used for cleaning, such as in public bath and shower facilities, IPX5 applies, which gives protection against water jets.

See {Table 2.4} for details of the index of protection (IP) system.

Current-using equipment may be installed in Zone 0 if it is fixed and permanently connected, is suitable for use in the zone according to manufacturer's instructions and is protected by SELV with a rated voltage not exceeding 12 V a.c. or 30 V d.c. Water heaters, showers and shower pumps may be installed in Zones 1 and 2, as may other equipments suitable for the conditions, provided that they are fixed and permanently connected, have the required protection against water and are protected by an RCD with a rating of not more than 30 mA.

Electric under-floor heating may be installed below any zone provided that a metallic sheath, enclosure or mesh connected to the protective system covers it. SELV under floor heating may be used in any zone provided that its safety source is installed outside Zones 0, 1 and 2, and is not required to have the earthed screening stated above. Table 8.1 summarises most of the requirements for equipment to be installed in the bathroom zones.

Table 8.1 Equipment to be installed in bathroom or shower room zones

Equipment	Zones				Comments
	0	1	2	outside	
LV light	√	√	√	√	Transformer outside zones 1 & 2
Mains light	x	x	√	√	
Pull switch	x	√	√	√	
Light switch	x	x	x	√	
Batten holder	x	x	x	√	
Whirlpool units	x	√	√	√	
Towel rails	x	√	√	√	
Mains fan	x	x	√	√	Must be to IP44
Shaver socket	x	x	√	√	Only if not subject to spray
Water heater	x	√	√	√	If suitable for the conditions
Electric shower	x	√	√	√	
Shower pump	x	√	√	√	If suitable for the conditions
SELV equip.	note	√	√	√	Transformer outside zones 1 & 2
Socket outlet	x	x	x	√	Only if 3.0 m from zone 1
Laundry equipment	x	x	x	√	Permanently connected, 30 mA RCD

Note Only 12 V SELV fixed equipment that cannot be located elsewhere.

8.3 SWIMMING POOLS AND OTHER BASINS

8.3.1 Introduction [702.11]

People using swimming pools, paddling pools and the basins of fountains are often partly unclothed and wet. The absence of clothing (particularly shoes) will remove much of their protection from shock (see {3.4}), and the water on their skin will tend to short circuit its natural protection. Thus, such people are very vulnerable to electric shock due to their reduced body resistance, so special measures are needed to ensure that the possibility of basic or fault contact (previously known as direct or indirect contact) is much reduced. Only separated extra-low voltage (SELV) or protective extra-low voltage (PELV) equipment may be installed, other than water heaters, in Zones 1 and 2, and equipment specially designed to be safe in the vicinity of swimming pools. Attention is drawn to the fact that the requirements of BS 7671:2008 do NOT apply to natural areas such as lakes, gravel pits and so on. Medical pools are also outside the scope of this Section, but see {8.19}.

Guidance Note 7 makes it clear that the basins of fountains that are intended to be occupied by persons (becoming quite common in some leisure applications) are subject to similar regulations as swimming pools; in fact the relevant Chapter in the note is titled 'Swimming Pools and Fountains'. Where it is not intended that people should occupy the basin of a fountain, requirements are relaxed as detailed in {8.3.2}.

The Regulations classify zones around the swimming or paddling pool, the arrangements being as follows:

Zone 0 is the inside of the pool or basin, including chutes and flumes, as well as apertures in the pool walls and floor which are accessible to the bathers, water jets or waterfalls and the space below them.

Zone 1 is a volume above the pool to 2.5 m above the rim, plus 2.5 m above the surrounding floor area on which people may walk, extending horizontally 2.0 m outwards from the rim. If the pool rim is above the surrounding floor level, the zone extends 2.5 m above the rim. Where the pool is provided with diving or spring boards, starting blocks or a chute, the zone also includes the volume enclosed by a vertical plane 1.5 m in front of and behind the ends of the board and extending upwards to 2.5 m above the highest surface which will be used by people, or to the ceiling if there is one.

Zone 2 is the volume extending 1.5 m horizontally from the boundary of Zone 1 and 2.5 m vertically above the floor. There is no zone 2 for a fountain.

The extents of the three zones are shown in Fig 8.3. Although limited by walls or fixed partitions the zones do extend through doorways and other openings.

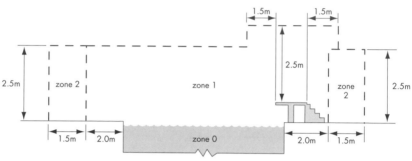

Fig 8.3 Dimensions of zones for swimming pools

8.3.2 Special requirements for swimming pools [702]

Note that the requirements for fountains differ somewhat – (see {8.3.3}).

Appliances and sockets for use in these high-risk areas must be separated extra-low voltage (SELV) type, at a potential not exceeding 12 V ac or 30 V dc, with the source being installed outside all three zones. Equipment intended for use in Zone 0 only when there are no people in that zone, for cleaning and other maintenance purposes, must be SELV or PELV with source outside the zones (although it can be in zone 2 if protected by a 30 mA RCD, or is itself protected by a 30 mA RCD). Socket outlets feeding such circuits must be provided with a notice indicating that they must only be used when there are no people in the pool.

Normal low voltage (230 V) luminaires are not permitted in zones 1 or 2, so general illumination is best provided by luminaires mounted higher than 2.5 m above floor level. Protection must be by electrical separation (see {5.8.4}) or by 30 mA residual current device(s). If RCDs are used, care must be taken to ensure that high leakage currents do not trip devices on starting; the use of multiple circuits, each with its own RCD, is advised. Attention is drawn to the difficulty of servicing luminaires mounted above the pool.

Luminaires for use under water or in contact with water in pools must be fixed and must comply with BS EN 60598. Where such lighting is installed behind water-tight portholes, and serviced from behind, it must comply with the same standard and there must be no possibility of contact between any part of the luminaire and any conductive part of the porthole. Electrical equipment in zones 0 and 1 must be mechanically protected by mesh, glass or grids that are only removeable by the use of tools. All luminaries in these zones must comply with BS EN 60598, and electric pumps with BS EN 60335.

Fixed SELV and PELV equipment designed for use in swimming pools, such as jet stream pumps and filtration systems, must meet all the following requirements:

1. regardless of the classification of the equipment, it must be located inside an insulated enclosure providing mechanical protection against impact of medium severity, and
2. it shall only be accessible via a hatch or door by means of a tool; opening the door must disconnect all live conductors, and
3. the supply circuit for the equipment must be protected by
   a) the SELV or PELV supplies at 25 V a.c. or 60 V d.c. must be from sources situated outside zones 0. 1 and 2, or
   b) protected by a 30 mA RCD, or
   c) having electrical separation [413], the source being outside zones 0 and 1.

In the case of a swimming pool that is too small to locate luminaries outside zone 1, they may be installed within the zone provided that they are more than 1.25 m (arm's reach) from the edge of zone 0 and also meet the three requirements for equipment immediately above.

In no case is it permissible to rely for protection against basic contact (previously known as direct contact) on obstacles, placing out of reach, a non-conducting location or earth-free equipotential bonding. If wiring is run on the surface it must not be metallic sheathed and must not be run in metal conduit or metal trunking, except in zone 2. Non-metallic cable enclosures are preferred.

Enclosures used for wiring systems or appliances at swimming pools are subject to special requirements, which depend on whether water jets will be used for cleaning purposes. If they will be used, protection must be to IPX8, (submersion in water), within the pool (Zone 0) and IPX5, which means that there must be effective protection against such water jets (see {Table 2.1}). If water jets will not be used, protection depends on the zone (see {8.3.1}) in which the system or appliance is situated. The requirements are:

Zone 0 IPX8, which means protection from submersion in water,

Zone 1 IPX4, so that protection is provided against splashing water or IPX5 where cleaning jets are used, and

Zone 2 indoor pools, IPX2, which gives protection against dripping water when inclined at 15°; and outdoor pools, IPX4, which gives protection against splashing water. In all cases, IPX5 applies where cleaning jets are used.

There must be no switchgear, control gear, socket outlets or accessories installed within zones 0 or 1, except for small pools where socket outlets are necessary and cannot possibly be installed outside zone 1. In this case, BS EN 60309 socket outlets may be installed provided they are at least 1.25 m (arms' reach) outside zone 0, are at least 0.3 m above the floor, and are protected by an RCD with a 30 mA rating or are protected by electrical separation [413] with the necessary isolating transformer situated outside zones 0, 1 and 2. Joint boxes may not be used in zones 0 or 1, other than on SELV and PELV systems. Instantaneous water heaters complying with BS 3456 may be installed in zones 1 and 2.

Socket outlets (in zone 2 only – but see above for small pools) must be to industrial standard BS EN 60309, and a shaver outlet to BS EN 60742 may also be installed in zone 2 (see {5.8.4}). If electric floor heating is used around a pool, it must have a metal sheath, or a covering metal grid if there is no metallic sheath, connected to the local equipotential bonding. Where the basin of a fountain is not intended to be occupied by people, installations in zones 0 and 1 may be protected by electrical separation (see {5.8.4}) or by automatic disconnection of the supply using an RCD with an operating current not exceeding 30 mA.

8.3.3 Fountains [702.55]

Fountains have no zone 2.

The basins of fountains are treated as swimming pools unless special means are taken to prevent people from gaining access to them. Where access is not prevented, protection must be by using SELV or PELV, with the source being outside zones 0 and 1, or by using an RCD rated at not more than 30 mA, or by electrical separation, each source of separation feeding only one item of equipment and installed outside zones 0 and 1.

Cables feeding equipment in zone 0 must be installed outside the basin (in zone 1) and such cables must be suitable for continuous immersion in water to a depth of 10 m. All electrical equipment in zones 0 and 1 must be made inaccessible by the use of meshes or grids. Luminaires mounted in these zones must be permanently fixed and must comply with BS EN 60598-2-18, and electric pumps with BS EN 60335-2-41. Fixed equipment for fountains fed from SELV and PELV supplies are required to meet the same three requirements as for swimming pools (see above).

8.4 SAUNA ROOMS

8.4.1 Introduction [703]

A sauna is a room in which the air is heated to a high temperature, humidity usually being very low, but occasionally increasing when water is deliberately poured over the heater. People using a sauna are usually unclothed and often wet (mainly due to perspiration), the absence of clothing (particularly shoes) removing much of their protection from shock (see {3.4.2}). If a sauna includes a cold basin or shower it is subject to the Regulations of the previous two sections {8.2 and 8.3}. As with bathrooms and swimming pools, zones are again classified, but this time they are concerned more with temperature levels than contact with water. The four zones (A, B, C and D) of the 16th Edition have been replaced in the 17th with three zones, 1, 2 and 3.

Zone 1 The volume within 0.5 m horizontally from the sauna heater, and extending from the floor up to the ceiling.

Zone 2 The volume covering the whole of the sauna room outside zone 1 up to 1.0 m above the floor.

Zone 3 The volume directly above zone 2 and extending upwards to the ceiling.

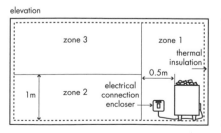 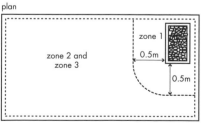

Fig 8.4 Definition of zones for sauna rooms

8.4.2 Special requirements for saunas [703.4, 703.5]

If separated extra-low voltage (SELV) is used, normal protection (formerly known as direct contact) must be provided by enclosures or barriers providing protection to IP24. This means protection against entry of human fingers and from splashing water.

Some of the standard measures for protection against both basic and fault protection (previously known as direct and indirect contact) must **not** be used in saunas. These are:

1 obstacles,
2 placing out of reach,
3 non-conducting location, and
4 earth free equipotential bonding.

30 mA RCDs must protect every circuit, either individually or collectively, other than for the sauna heater itself unless recommended by the manufacturer. Attention is drawn to the high leakage currents common with sauna heating elements, particularly when water is applied. The manufacturer should be consulted to ensure that leakage currents are unlikely to cause problems with an RCD rated at 30 mA. In some cases an RCD with a 100 mA rating may be more satisfactory. All equipment must be protected to at least IPX4 or IPX5 where water jets are used and no equipment other than the sauna heater, which must comply with BS EN 60335-2-53 and must be installed in accordance with the manufacturer's instructions, may be installed in zone 1. There must be no socket outlets in a sauna room, nor other switchgear that is not built into the sauna heater; lighting switches must be positioned outside the sauna room. As well as having no socket outlets within the sauna room, it is advisable not to install them close to the room at all, which could encourage the introduction of portable appliances. In the lower part of the room where it will not be so hot (zone 2) there is no special requirement concerning the heat resistance of equipment. If installed in zone 3, equipment must be suitable for operation at an ambient temperature of 125°C, with the sheaths of cables able to withstand temperatures of 170°C. The sauna room is likely to be provided with thermal insulation, and wiring should ideally be external, so that temperature requirements for cables will not apply.

8.5 INSTALLATIONS ON CONSTRUCTION AND DEMOLITION SITES

8.5.1 Introduction [704]

The electrical installation on a construction or demolition site is there to provide lighting and power to enable the work to proceed. By the very nature of the situation, the installation will be subjected to the kind of ill treatment that is unlikely to be applied to most fixed installations. Those working on the site may be ankle deep in mud and thus particularly susceptible to a shock to earth, and they may be using portable tools such as drills and grinders in situations where danger is more likely than in most factory situations. Where work takes place in cramped and confined locations, reference should be made to [706], 'Conducting locations with restricted movement', covered here in {8.7}. The difficulty of ensuring that bonding requirements are met on construction sites means that PME supplies must not be used to supply them.

The 16th Edition requirement of lower earth fault loop impedance values to give reduced disconnection times has been dropped from the 17th Edition, so the values given in {Tables 5.1 and 5.2} will apply to these installations.

These installations will also, by definition, be temporary. As the construction proceeds they will be moved and altered. It is usual for such installations to be subjected to thorough inspection and testing at intervals which will never exceed three months. A formal visual inspection of 110 V equipment should take place monthly; the effectiveness of each RCD should be tested before every use by pressing its test button. BS 7375 is a useful source of information.

As well as the erection of new buildings, the requirements for construction sites will also apply to:

1 sites where repairs, alterations or additions are carried out
2 demolition of buildings
3 public engineering works
4 civil engineering operations, such as road building, coastal protection, earthworks, etc.

The special requirements for construction sites do not apply to temporary buildings erected for the use of the construction workers, such as offices, toilets, cloakrooms, dormitories, canteens, meeting rooms, etc. Neither do they apply to open cast mining operations. These situations will not change as construction progresses, and are thus subject to the general requirements of the Regulations.

The equipment used must be suitable for the particular supply to which it is connected, and for the duty it will meet on site. Where more than one voltage is in use, plugs and sockets must be non-interchangeable to prevent misconnection. Hand lamps on construction sites must now be fed from SELV supplies, the requirements for which will be found in {8.17.2}. The five voltage levels acceptable for construction sites are:

1 400 V three phase, for use with fixed or transportable equipment with a load of more than 3750 W,
2 230 V single phase, for site buildings and fixed lighting,
3 110 V three phase star point earthed, for transportable equipment with a load up to 3750 W, reduced low voltage systems, portable hand-held tools and local lighting up to 2 kW.
4 110 V single phase, fed from a transformer, often with an earthed centre-tapped secondary winding, to feed transportable tools and equipment, such as floodlighting with a load of up to 2 kW and portable hand-lamps for general use. This supply ensures that the voltage to earth should never exceed 55 V (see {Fig 8.5}). The primary winding of the transformer must be RCD-protected.
5 SELV for portable hand-lamps in confined or damp situations.

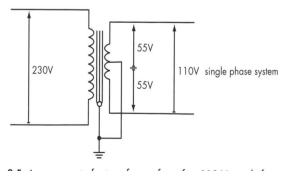

Fig 8.5 Arrangement of a transformer for safety 110 V supply for a construction site

Supplies will normally be obtained from the Electricity Supply Company. Where a site is remote, so that a generator must be used (IT supply system) special protective requirements apply which are beyond the scope of this Guide, and the advice of a qualified electrical engineer must be sought. Attention is drawn to BS 4363: 1998 Specification for distribution assemblies for electricity supplies for construction and building sites.

8.5.2 Special regulations for construction sites [704.3 to 704.5]

Construction site installations are like most others in that they usually rely on earthed equipotential bonding and automatic disconnection for protecting from electric shock. This is the system where an earth fault, which results in metalwork open to touch becoming live, also causes a fault current which will open the protective device to remove the supply within 0.4 s for socket outlet circuits or 5 s for fixed appliances. Where values of maximum earth-fault loop impedance are necessary to check compliance with this requirement, the values of {Tables 5.1 and 5.2} apply. Where the given values of earth-fault loop impedance cannot be met, protection must be by means of RCDs with operating current not exceeding 30 mA. Protection by obstacles or barriers is not acceptable except for SELV and PELV systems.

For fixed installations (limited to main switchgear and principal protective devices) the disconnection time is 5 s, so {Tables 5.2 and 5.4} can be used to find maximum earth-fault loop impedance values.

Sockets on a construction site must be safety extra-low voltage (SELV) or protected by a residual current circuit breaker (RCD) with an operating current of not more than 30 mA, or must be electrically separate from the rest of the supply, each socket being fed by its own individual transformer. SELV is unlikely for most applications, because 12 V power tools would draw too much current to be practical. Most sockets are likely to be fed at 110 V from centre-tapped transformers (see {Fig 8.5}) so will comply with this requirement; they must be protected by an RCD with a rating of 500 mA.

Distribution and supply equipment must comply with BS 4363, and, together with the installation itself, must be protected to IP44. This means provision of mechanical protection from objects more than 1 mm thick and protection from splashing water. Such equipment will include switches and isolators to control circuits and to isolate the incoming supply. The main isolator must be capable of being locked or otherwise secured in the 'off' position. Emergency switches should disconnect all live conductors including the neutral.

Cables and their connections must not be subjected to strain, and cables must not be run across roads or walkways without mechanical protection. Where feeding reduced low voltage equipments, thermoplastic (pvc) flexible cables may be used, but for higher voltages the flexible cables must be H07RN-F type or equivalent having 450/750 V rating and being resistant to abrasion and water. Circuits supplying equipment must be fed from a distribution assembly including overcurrent protection, a local RCD if necessary, and socket outlets where needed. Socket outlets must be enclosed in distribution assemblies, fixed to the outside of the assembly enclosure, or fixed to a vertical wall. Sockets must not be left unattached, as is often the case on construction sites. Socket outlets and cable couplers must be to BS EN 60309-2. Luminaire supporting couplers must not be used. A typical schematic diagram for a construction site system is shown in {Fig 8.6}.

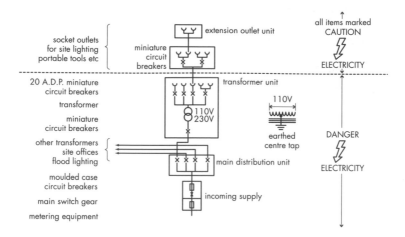

Fig 8.6 Single-phase distribution system for a construction site to BS 4363 and CP1017

8.6 AGRICULTURAL AND HORTICULTURAL INSTALLATIONS

8.6.1 Introduction [705]

Situations of these kinds will pose an increased risk of electric shock because people and animals are more likely to be in better contact with earth than in other installations. Animals are particularly vulnerable, because their body resistance is much lower than for humans, and applied voltage levels which are quite safe for people may well prove fatal for them. If animals are present, the electrical installation may be subjected to mechanical damage (animals cannot be instructed to keep away from installations) as well as to a greater corrosion hazard due to the presence of animal effluents. The 16th Edition requirement of lower earth fault loop impedance values to give reduced disconnection times has been dropped from the 17th Edition, so the values given in {Tables 5.1 and 5.2} will apply to these installations. TN-C systems (earthed concentric wiring) are not permitted in agricultural and horticultural situations.

The special requirements for agricultural and horticultural installations also apply to locations where livestock are kept, such as stables, chicken houses, piggeries, etc. They also apply to feed processing locations, lofts, and storage places for hay, straw and fertilizers.

For these reasons, special Regulations apply to such installations, both indoors and outdoors. It is important to recognise that dwelling houses on agricultural and horticultural premises used solely for human habitation are excluded from these special requirements.

8.6.2 Agricultural installations [705.4, 705.5]

All systems, including as those complying with SELV requirements, must be protected to IP44 (i.e. protected from solid objects not exceeding 1 mm and from splashing water).

All socket outlets rated at 32 A or less must be protected by residual current device(s) (RCDs) with an operating current of no more than 30 mA, whilst for those rated at over 32 A a 100 mA RCD is required. Socket outlets in agricultural and horticultural premises must be:

1 to BS EN 60309-1, or
2 to BS EN 60309-2 when interchangeability is required, or
3 to BS 1363, BS 546 or BS 196 provided that the rated current does not exceed 20 A.

Sockets must not be situated close to combustible materials. It is accepted that earthed equipotential bonding and automatic disconnection (sometimes known as EEBAD) cannot protect livestock because the voltages to which they would be subjected in the event of a fault are unsafe for them. Fixed equipment and distribution circuits are permitted to have a disconnection time of 5 s, which is the same as for normal installations, and require the maximum earth-fault loop impedance values shown in {Tables 5.2 and 5.4}.

Supplementary equipotential bonding must be applied to connect together all exposed and extraneous conductive parts that are accessible to livestock and the main protective system. It is recommended that a metallic grid should be laid in the floor and connected to the protective conductor. Bonding must use either 4 mm^2 copper or hot galvanised steel conductors, the latter being either flat 30 mm x 3 mm or circular 8 mm in diameter. Care must be taken to protect the bonding system from both chemical and electrolytic corrosion.

Fire is a particular hazard in agricultural premises where there may be large quantities of loose straw or other flammable material. A particular fire hazard on agricultural premises is caused by damage to the wiring by rodents gnawing at cables. This effect can be reduced by cable runs that are below ceilings rather than in roof spaces and by the use of steel conduits. The Regulations require the protection of the system by an RCD with an operating current not greater than 300 mA. This will result in problems of discrimination between this unit and those of lower operating current rating unless the main RCD is of the time delayed type (see {5.9.3}). Care must be taken to ensure that heaters are not in positions where they will ignite their surroundings; a clearance of at least 500 mm is required for radiant heaters (see also {3.9.4}). Where electric heating is used for breeding and rearing situations, it must comply with IEC 60335-2-71. In high-density livestock rearing situations, life support must be provided by:

1 a backup supply where needed for food, water, ventilation and lighting
2 discrimination to ensure the continuation of ventilation
3 a standby source or temperature and supply voltage monitoring for ventilation.

All electrical equipment must be protected to IP44, and chosen to be suitable to operate under the onerous conditions they will experience. Wiring and other electrical equipment must be inaccessible to livestock and must be vermin proof. In practice, this will probably mean enclosure in galvanised steel conduit, or the use of mineral insulated cables. Where conduit is not used, cables to H07RN-F should be considered. Switch and control gear must be to IP44 and constructed of, or enclosed in, insulating material. Switches for emergency use must not be in positions accessible to cattle, or where cattle may make operation difficult. Emergency switches should disconnect all live conductors including the neutral. It may be necessary to omit isolators in some cases to ensure that essential supplies (such as broiler house fans) are not disconnected unintentionally. The likelihood of panic amongst animals when emergencies occur must be taken into account. Luminaires must be to IP54 and suitable for mounting on a flammable surface.

Where cables are buried the depth must be at least 0.6 m with covering protection (cable tiles), increased to 1.0 m where the ground is cultivated. Overhead lines must be insulated and at least 6.0 m above ground level.

After testing, the user of the installation must be provided with documentation including:-

1 a plan to show the location of all electrical equipment
2 drawings to show the routing of all concealed cables
3 a single-line distribution diagram
4 an equipotential bonding diagram showing the locations of bonding connections.

8.6.3 Electric fence controllers

Unlike its predecessor, the 17th Edition makes no mention of supplies for electric fences. However, they are in wide use and it seems sensible to follow the guidance of the 16th Edition. Electric fences are widely used to prevent animals from straying. In most cases they are fed with very short duration pulses of a voltage up to 5 kV - however, the energy involved is too small to cause dangerous shock. Usually animals very quickly learn that it can be painful to touch a fence, and give it a wide berth. {Fig 8.7} indicates a typical arrangement for an electric fence, the numbers in brackets in the following text relating to the numbers on the diagram.

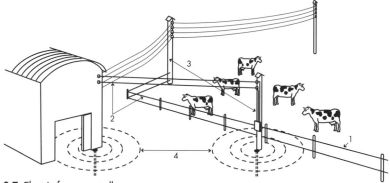

Fig 8.7 Electric fence controller

Controllers may be mains fed or battery operated. Controllers must be to BS EN 61011-1, and not more than one controller may be connected to a fence (1). So that it will give a short, sharp shock to animals, the high voltage output of the controller is connected between the fence and an earth electrode. It is of obvious importance that the high voltage pulses are not transmitted to the earthed system of a normal electrical installation because the earth zones overlap (4). A method of checking the earth zone of an electrode is described in {7.6.1}. For similar reasons it is important that the fence should never make contact with other wiring systems (2), telephone circuits, radio aerials, etc; the installer must also take account of induction from overhead lines, which may occur when the fence runs for a significant distance below such a line.

A mains operated fence controller must be installed so that interference by unauthorised persons and mechanical damage is unlikely. It must not be fixed to a pole carrying mains or telecommunication circuits, except that where it is fed by an overhead line consisting of insulated conductors, it may be fixed to the pole carrying the supply (3).

8.6.4 Horticultural installations [705]

A horticultural installation is likely to be subject to the same wet and high-earth-contact conditions experienced in agriculture, but will be free from the extra hazards associated with the presence of animals. In other respects it is subject to the same requirements as an agricultural installation as described in {8.6.2}. A type of load not so likely to be present in the agricultural situation is the soil warming circuit often used in horticulture. The relevant Regulations for such circuits are considered in {8.12.4}.

8.7 RESTRICTIVE CONDUCTIVE LOCATIONS

8.7.1 Introduction [706]

A restrictive conductive location is one in which the surroundings consist mainly of metallic or conductive parts, with which a person within the location is likely to come into contact with a substantial portion of his body, and where it is very difficult to interrupt such contact. An example could be a metal enclosure, such as a boiler, inside which work must be carried out. Clearly, such a situation could result in considerable danger to a worker, who may be lying inside the boiler whilst using an electric drill or grinder.

8.7.2 Special requirements [706.4]

The person in the restrictive location is protected by the use of the separated extra-low voltage (SELV) system. A SELV supply must be fed from a safety source, such as a special step-down isolating transformer. Protection against fault contact by obstacles or placing out of reach is not permitted.

Protection against fault contact (touching parts which have become live as the result of a fault) for fixed equipment is achieved by one of the following methods.

1 For the supply to a hand-held tool or to portable equipment
    a) electrical separation [413], but only one item may be connected to the secondary winding of a transformer. Note, however, that the transformer may have a more than one secondary winding
    b) SELV as [414] and {8.17.2}
2 For the supply to hand lamps, likely to be essential within an enclosure, SELV as [414]. Note that a fluorescent lamp with a built-in step-up transformer is permissible provided that it has electrically separated windings
3 For the supply of fixed equipment
    a) automatic disconnection of the supply [411] with supplementary equipotential bonding [415.2], or
    b) Class II equipment with the additional protection of a 30 mA RCD, or
    c) electrical separation with only one item of equipment being connected to the secondary of the isolating transformer, or
    d) SELV, or
    e) PELV, see {8.17.3}, which is connected to earth and equipotential bonding is provided between all exposed conductive parts, all extraneous conductive parts within the location and the connection of the PELV system to earth.

Separated and isolating sources, other than those specified above, must be situated outside the restrictive conductive location unless they are part of an installation in a permanent restrictive conductive location. By its very nature an enclosed space is likely to be dark, and handlamps will be needed by the operator. Socket outlets intended to feed handlamps must be SELV, and these outlets can also be used to feed hand tools. However, the limitation of 25 V on SELV means that the range of hand tools available is very limited, and where higher voltages are used, sockets for this purpose can be fed using electrical separation.

8.8 EARTHING FOR INSTALLATIONS HAVING HIGH PROTECTIVE CONDUCTOR CURRENTS

8.8.1 Introduction [543.7]

In the 16th Edition of BS 7671 these requirements were included under 'Special Installations and Locations'. In the 17th they have been moved to the main body of the Regulations as [543]. Data processing equipment is extremely common. Such computer-based equipments are used in the office for word processing, handling accounts, dealing with wages and so on. In the factory they control processes, and in the retail shop they are used for stock control, till management and many other purposes. Even in the home they are used for security, for temperature control, for domestic banking services, etc. Telecommunications equipment is also very widely used.

All such equipments have a common danger of failing and losing their stored data if subject to mains disturbances such as voltage spikes and transients. They are protected from such failures by feeding the supplies to them via filter circuits, which are designed to remove or reduce such voltage variations before they reach the sensitive circuitry. A simple filter is shown in {Fig 8.8} and will almost always include resistive and capacitive components, which are connected from live conductors to earth. This will give rise to increased earth currents, driven by supply voltage through the resistive and reactive components. When the circuit, including the filter, is switched on, higher earth currents will usually flow for a very short time whilst capacitors are charging. Many modern luminaries, using electronic control rather than the older inductive choke, also draw very high protective conductor currents.

Data processing and telecommunications equipment of this kind therefore has high levels of earth current in normal use, although in most cases the earth current produced by a filter will not exceed 3.5 mA. Where a number of equipments containing filters are fed from a single circuit, the total supply current may be low due to the small demand of each device, but the total earth leakage current may well be very high. Consequently, special regulations apply to the protective and earthing conductors of such circuits.

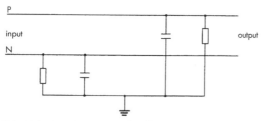

Fig 8.8 Mains transient suppression filter

8.8.2 Special regulations for equipment with high protective conductor currents [543.7.1]

Electricity Supply Regulation 26 indicates that the level of earth leakage current should not normally exceed one ten thousandth part of the installation maximum demand (for example, 10 mA earth leakage current for an installation with a maximum demand of 100 A). Data processing equipment, apparatus having radio frequency interference suppression filters and heating elements, are likely to have a higher leakage current than this, so special regulations become necessary. Foremost is the requirement that where earthing is used for functional purposes (to allow the filters to do their job) as well as protective purposes, the protective function must take precedence. When the earth leakage current is high, serious shocks are likely from accessible conductive parts which are connected to a protective conductor which is not itself solidly connected to the main earth terminal.

Where a distribution board feeds outlets for high leakage current equipment, information must be added to the board data to draw attention to such equipment.

Stationary equipment with an earth leakage current exceeding 3.5 mA must preferably be permanently connected, or an industrial plug and socket to BS EN 60309-2. When a socket outlet circuit may be expected to feed data processing equipment with normal earth leakage current of more than 10 mA, or if the earth leakage current for a circuit feeding fixed stationary equipment is greater than 10 mA, earthing must be through a high integrity protective system complying with at least one of the following:

1 a protective conductor of cross-sectional area at least 10 mm² for permanently connected equipment, or 4 mm² where additionally protected from mechanical damage

2 duplicated protective conductors having separate connections

3 duplicated protective cross-sectional areas of all the conductors is at least 10 mm², in which case the metallic sheath, armour or braid of the cable may be one of the protective conductors, provided that it complies with the adiabatic equation (see {5.1.5})

4 duplicated protective conductors, one of which can be metal conduit, trunking or ducting, whilst the other is a 2.5 mm² conductor installed within it

5 an earth monitoring device is used which switches off the supply automatically if the protective conductor continuity fails

6 the supply through a double wound transformer, a protective conductor complying with one of the arrangements 1 – 5 above connecting exposed conductive parts to a point on the secondary winding.

7 where an industrial plug and socket is used, the protective conductor in the flexible cable or cord must be at least 2.5 mm² for 16 A plugs and 4 mm² for those of higher rating

The reason for these precautions is that if the circuit protective conductor should become open circuit, leakage current could flow to earth through a person touching exposed conductive parts with possibly lethal consequences.

The alternative to items 1 to 7 above is the use 13 A sockets to BS 1363 wired on a ring or radial circuit. In all cases the protective conductors must have a minimum size of 1.5 mm² with both ends of the ring connected to separate earth terminals at the distribution board. There must be no spurs on ring circuits, or branches on radial circuits, which must have a duplicate protective conductor as shown in {Fig 8.9}. Spurs are permitted from 30 A or 32 A busbar systems which are often used for supplying IT equipment. In such cases equipment with a protective conductor current that exceeds 3.5 mA in normal service must not be used. Where twin or triple socket outlets are used, they must be provided with two earth terminals, one for each protective conductor. Where protective conductors are not in the same sheath, conduit or trunking as circuit conductors, they must be 4 mm² rather than 1.5 mm².

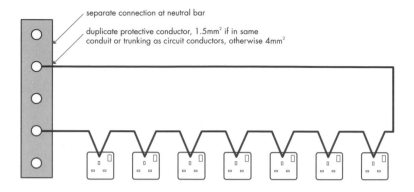

separate connection at neutral bar

duplicate protective conductor, 1.5mm² if in same conduit or trunking as circuit conductors, otherwise 4mm²

Fig 8.9 Radial circuit with duplicated protective conductor

8.9 CARAVAN AND CAMPING PARKS AND SIMILAR LOCATIONS

8.9.1 Introduction [708]

The 16th Edition of BS 7671 had one Section [608] to deal with caravans, the parks on which they are situated, and all related matters. The 17th Edition has a changed approach, [708] covering the park and the overhead and underground supplies to the vehicles on it and [7.21], dealing with the Regulations covering the caravans and motor caravans themselves. This volume follows a similar path, this Section dealing with the parks and {8.10} with the caravans and motor caravans. It does not apply to the installations of mobile or transportable units (see {8.25}).

The supply voltage must not be greater than 230 V single phase or 400 V three phase.

8.9.2 Leisure vehicle park installations [708.4, 708.5]

These Regulations cover the arrangements for the supply of electrical energy to individual plugs on a caravan or tent park. The plots on which caravans will stand are referred to as caravan pitches.

The supply to caravans may be PME, but must not provide an earthing connection - this means that only TT or TN-S systems may be used. If the supply system is PME, there must be separation of the earthing systems of the pitch supplies and the incoming supply, which is usually best effected at the main distribution board with a separate earth electrode for the pitch supplies. The PME system may be used to supply permanent buildings such as toilet blocks.

Wherever possible, the supplies to the plugs should be by means of underground cables (see {8.14.3}). These cables should be installed at least 0.6 m deep outside the area of the caravan pitch to ensure that they are not damaged by tent pegs (which are often used in erecting caravan awnings) and ground anchors which are used to secure the caravans against the effects of high winds.

If overhead supplies are used they must:

1 be constructed with insulated, rather than bare, cables
2 always be at least 2 m outside the area of every pitch
3 have a mounting height of at least 3.5 m, which must be increased to 6 m

where vehicle movements are possible. Since in most cases the whole of the area of a caravan site is subject to vehicle movements, most overhead systems will need to be at a minimum height of 6 m. Poles or other supports for overhead wiring must be located or protected so that they are unlikely to be damaged by vehicle movement.

There must be at least one supply socket for each pitch, the socket positioned so that it is between 0.5 m and 1.5 m above ground level (in special cases such as the risk of flooding or heavy snowfall this height may be exceeded), and it must be situated not more than 20 m from the caravan or tent. No more than four outlets may be grouped together to prevent supply cables crossing pitches other than the one they are intended to supply. The current rating must be at least 16 A, and the outlet must be to BS EN 60309-2, of the splash proof type to IPX4, and with the keyway at position 6h. If the expected current demand for a pitch exceeds 16 A, extra sockets of higher rating must be installed. A problem frequently arises due to the increasing use of 3 kW instantaneous water heaters in caravans because supply systems have not usually been designed to allow for such heavy loading.

Each socket must have its own individual overcurrent protection in the form of a fuse or circuit breaker; to ensure that enough current will flow to open the protective device in the event of an earth fault, it is recommended that the earth-fault loop impedance at the plug should not exceed 2 Ω. Every socket must be individually protected by an RCD complying with BS EN 61008-1 or BS EN 61009-1 with a 30 mA rating. Since a 30 mA RCD is also required within each caravan, the device protecting the pitch socket(s) should be time delayed. Equipment installed outdoors in caravan parks must be protected to IP34.

8.10 ELECTRICAL INSTALLATIONS IN CARAVANS AND MOTOR CARAVANS

8.10.1 Introduction [721]

A caravan is a leisure accommodation vehicle that reaches its site by being towed by another vehicle. A motor caravan is used for the same purpose, but has an engine, which allows it to be driven; the accommodation module on a motor caravan may sometimes be removed from the chassis. Caravans and motor caravans will often contain a bath or a shower, and in these cases the special requirements for such installations (see {8.2}) will apply. Railway rolling stock is not included in the definition as a caravan. Caravans used as mobile workshops will be subject to the requirements of the Electricity at Work Regulations 1989 as well as BS 7671: 2008 and see {8.25}. These Regulations do not apply to the automotive aspects of a motor caravan, but see [Annex A to 721] concerning requirements such as those for direction indicators and rear lights on the caravan whilst it is under tow.

All the dangers associated with fixed electrical installations are also present in and around caravans. Added to these are the problems of moving the caravan, including connection and disconnection to and from the supply, often by totally unskilled people. Earthing is of prime importance because the dangers of shock are greater. For example, the loss of the main protective conductor and a fault to the metalwork in the caravan is likely to go unnoticed until someone makes contact with the caravan whilst standing outside it {Fig 8.10}. The requirements of the Electricity Supply Regulations do not allow the supply neutral to be connected to any metalwork in a caravan, which means that PME supplies must not be used to supply them.

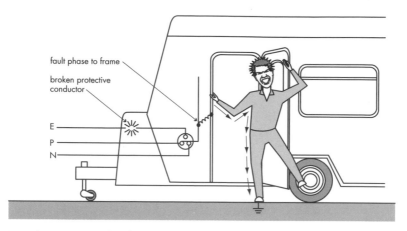

Fig 8.10 The importance of earthing a caravan

8.10.2 Caravan installations [721.31 to 721.55]

Note that motor caravans are subject to the same regulations as caravans. Except for a shaver socket, the protective measure of electrical separation is not permitted. Neither are the measures of obstacles, placing out of reach, non-conducting location and earth-free local equipotential bonding. The requirements for the electrical installation are listed below, numbers relating to those shown on {Fig 8.11}.

1 An inlet coupler to BS EN 60309-2 with its keyway at position 6h must be provided no higher than 1.8 m above the ground.

2 A spring hinged lid which will close to protect the coupler socket when travelling must be fitted.

3 A clear and durable notice must be provided beside the inlet coupler to indicate the voltage, frequency and current of the caravan installation.

4 Protection by automatic disconnection (as for all other installations) but using double pole MCBs to disconnect all live conductors, together with a double pole RCD complying with BS EN 61008-1 or BS EN 61009-1 with an operating current of 30 mA and means of isolating the complete caravan installation must be used.

Table 8.5 Instructions for electricity supply

To connect

1 Before connecting the caravan installation to the mains supply, check that:
   a) the supply available at the caravan pitch supply point is suitable for the caravan installation and appliances, and
   b) the voltage and frequency and current ratings are suitable, and
   c) the caravan main switch is in the off position

2 Open the cover to the appliance inlet provided at the caravan supply point and insert the connector of the supply flexible cable

3 Raise the cover from the electricity outlet provided on the pitch supply point and insert the plug of the supply cable.

The caravan supply flexible cable must be fully uncoiled to avoid damage by overheating

4 Switch ON at the caravan main switch

5 Check the operation of residual current devices, if any, fitted in the caravan by depressing the test button.

In case of doubt or if, after carrying out the above procedure the supply does not become available, or if the supply fails, consult the caravan park operator or his agent or a qualified electrician

To disconnect
6   Switch off at the caravan main isolating switch and unplug both ends of the cable.

Periodical inspection
Preferably not less than once every three years and more frequently if the vehicle is used more than the normal average mileage for such vehicles, the caravan electrical installation and supply cable should be inspected and tested and a report on its condition obtained as prescribed in BS 7671 (formerly the Regulations for Electrical Installations) published by the Institution of Electrical Engineers.

5   A durable notice must be fixed beside the isolator with the wording shown in {Table 8.5}.

6   All circuits must be provided with a protective conductor. All sockets are to be three pin with earthed contact and must have no accessible conductive parts. If two or more systems at different voltages are in use, the plugs of the differing systems must not be interchangeable. ELV sources must be 12 V, 24 V or 48 V when d.c., or 12 V, 24 V, 42 V or 48 V when a.c. Such ELV sockets must have their voltage clearly marked. Wiring may be insulated single-core flexible conductors or insulated single-core conductors with at least seven strands, both in non-metallic conduit (wiring systems which will allow fire to spread along them must not be used) or sheathed flexible cables. All conductors must have a cross-sectional area of at least 1.5 mm². Where 230 V and extra-low voltage circuits (usually 12 V) are both used, the cables of the two systems must be run separately and must both be insulated for 230 V. Since the wiring will be subjected to vibration when the caravan is moved, great care must be taken to ensure that bushes or grommets are used where it passes through metalwork.

7   Enclosed luminaires must be fixed directly to the structure or lining of the caravan. Where luminaires are of the dual voltage type (230 V mains and 12 V battery supply) they must be fitted with separate and different types of lampholder, with proper separation between wiring of the two supplies, and be clearly marked to indicate the lamp wattages and voltages. They must be designed so that both lamps can be illuminated at the same time without causing damage.

8   Pendant luminaires must have arrangements to secure them whilst the caravan is being moved.

9   All metal parts of the caravan, with the exception of metal sheets forming part of the structure, must be bonded together and to a circuit protective conductor, which must not be smaller than 4 mm² except where it forms part of a sheathed cable or is enclosed in conduit.

10 Sheathed cables must be supported at intervals of at least 400 mm where run vertically, or 250 mm horizontally unless run in non-metallic rigid conduit.

11 No electrical equipment may be installed in a compartment intended for the storage of gas cylinders.

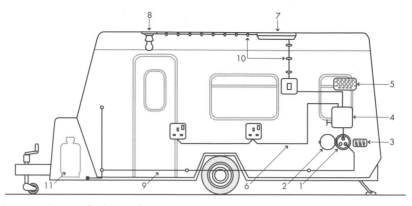

Fig 8.11 Requirements for the installation in a caravan

Where a caravan appliance may be exposed to the effects of moisture, it must be protected to IP55 (protected from dust and from water jets). Every caravan that includes an electrical installation must be provided with a flexible lead 25 m long fitted with a BS EN 60309-2 plug and a BS EN 60309-2 connector with the keyway at position 6h. The cross-sectional area of the cable must be related to the rated current of the caravan and the plug as shown by {Table 8.6}.

Table 8.6 Cross-sectional areas of flexible cables and cords for supplying caravan connectors from [Table 721] of BS 7671:2008)

Rated current (A)	Minimum cross-sectional area (mm²)
16	2.5
25	4.0
32	6.0
63	16.0
100	35.0

8.10.3 Tent installations

BS 7671 does not provide specific requirements for electrical installations in tents. However, such installations are becoming common, especially in permanently erected tents offered by holiday firms, the supply being derived from the park installation. For such installations, and for installations in trailer tents, it is recommended that the requirements for caravans stated earlier in this section should apply. See also {8.20} referring to shows and stands.

8.11 HIGHWAY POWER SUPPLIES AND STREET FURNITURE

8.11.1 Introduction [559.10]

In the 16th Edition of BS 7671 these requirements were included under 'Special Installations and Locations'. In the 17th they have been moved to the main body of the Regulations as [559.10]. The need for these regulations has arisen because of the increasing use of electrical supplies for street and footpath lighting, traffic signs, traffic control and traffic surveillance equipment. These uses are described as street furniture. Also covered by the Regulations is street located equipment that includes telephone kiosks, bus shelters, advertising signs and car park ticket dispensers. As far as the Regulations are concerned, no distinction is drawn between the two categories, both street furniture and street located equipment being subject to their requirements.

It must not be thought that these Regulations apply only to installations on public highways. They must also be followed in installations within private roads, car parks, and other areas where people will be present. They do not apply to suppliers' works, which in this context means the overhead or underground supplies feeding the equipment. It does apply, however, to overhead or underground supplies installed to interconnect street furniture or street located equipment from the point of supply.

8.11.2 Highway and street furniture regulations [559.10]

The equipment covered by these regulations is always, by definition, accessible to people of all kinds; however, maintenance must only be carried out by skilled persons. Thus, certain measures for protection in areas only open to skilled persons are not appropriate here. For example, protection by barriers or by placing out of reach must not be used, because those using the areas may not be aware of the dangers that follow from climbing over barriers or reaching up to normally untouchable parts. The equipment considered here usually has doors that give access to the live parts inside. Such doors, unless more than 2.5 m above ground level, must not allow unauthorised access to live parts, and must therefore either:

1 be opened only with a special key or tool, or

2 be arranged so that opening the door disconnects the live parts.

The doors of street furniture, such as lamp standards, are often damaged or even totally removed. Therefore, live parts within must be protected by enclosures or barriers to give protection to IP2X, to ensure that they cannot be touched. All cables buried directly should have a marker tape placed above them, 150 mm below the ground surface. To prevent disturbance, burial depths are usually 450 mm below verges and 750 mm below the highway. Very careful plans and records must be kept to show exactly where and how deeply such cables are buried.

The danger to the people who may touch metalwork is no more in this case than with electrical equipment indoors, so a 5 s disconnection time is acceptable where earthed equipotential bonding and automatic disconnection of the supply is employed. Installations of this type are usually simple, often consisting of a single circuit feeding lighting. Provided that there are no more than two circuits, there is no need to provide a main switch or isolator, the supply cut-out (main supply fuse in most cases) being used for this purpose if not rated at more than 16 A, but only by skilled or instructed persons. However, where the supply is provided by a separate supplier (the Electricity Supply Company), their consent to use of the cut-out for this purpose must first be obtained as with any other installation.

Internal wiring in all street electrical equipment must comply with the normal Regulations concerning protection, identification and support (see {4.4.1, 4.5, and 4.6}). All electrical enclosures for highway power supply, street furniture and street located equipment must have protection to IP33 (see {Section 2.4.3}). Attention is drawn particularly to the need to support cables in vertical drops against undue stress. The notice indicating the need to provide periodic testing is unnecessary where an installation is subject to a planned inspection and test routine.

Temporary installations such as those for Christmas or summer external lighting schemes are usually connected to highway power supplies. In many cases, street furniture is equipped with temporary supply units from which such installations can be fed. The temporary supply unit must have a clear external label indicating the maximum current it is intended to supply. Attention is drawn to the possibility of damage to existing cable connections by the frequent connection and disconnection of temporary supplies. It is recommended that a socket outlet, especially intended to feed temporary installations, should be part of the temporary supply unit and should be fixed within the street furniture enclosure. Such installations must comply with the requirements for construction site installations (see {8.5}) and must not reduce the safety of the existing installation.

Highway equipment is usually subjected to vibration, corrosion and condensation, and sometimes to vandalism. It should be chosen with such problems in mind. The heat produced by lamps and control gear will usually be sufficient to prevent condensation and corrosion, but thought should be given to the provision of a low power heater in other cases.

Inspection and testing is necessary as for all electrical installations, and should be synchronised with other maintenance work, such as white lining and re-lamping, to avoid inconvenience to highway users as far as is possible. In general, a period of six years between tests is acceptable for fixed installations, and three months for temporary systems.

8.12 HEATING APPLIANCES AND INSTALLATIONS
8.12.1 Introduction
This section is concerned with the special requirements for devices and circuits that are designed to produce heat for transfer to their surroundings. All water heaters (as well as those for other liquids) must be provided with a thermostat or cutout to prevent a dangerous rise in temperature. Not only could the high temperature of the water or other liquid be dangerous, but if allowed to boil, very high pressures, leading to the danger of an explosion, could result if the liquid container were sealed.

8.12.2 Electrode boilers and water heaters [554.1]
An electrode heater or boiler is a device that heats the water contained, or raises steam.

Two or three electrodes are immersed in the water and a single- or three-phase supply is connected to them. There is no element, the water being heated by the current that flows through it between the electrodes. Electrode heaters and boilers must be used on a.c. supplies only, or electrolysis will occur, breaking down the water into its components of hydrogen and oxygen. The different requirements for single-phase and three-phase boilers are shown below.

A **Three-phase electrode heater fed from a low voltage supply**

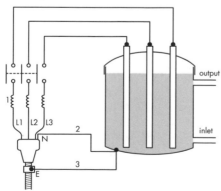

Fig 8.12 Three-phase electrode boiler fed from a low voltage supply

The requirements are:

1 a controlling circuit breaker which opens all three phase conductors and is provided with protective overloads in each line,

2 the shell of the heater bonded to the sheath and/or armour of the supply cable with a conductor of cross-sectional area at least equal to that of each phase conductor,

3 the shell of the heater bonded to the neutral by a conductor of cross-sectional area at least equal to that of each phase conductor.

B Three-phase electrode heater fed from a supply exceeding low voltage

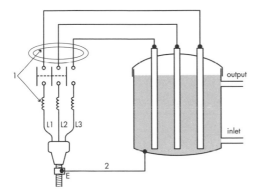

Fig 8.13 Three-phase electrode boiler fed from a supply exceeding low voltage

The requirements are:

1 a controlling circuit breaker fitted with residual current tripping, set to operate when the residual current is sustained and exceeds 10% of the rated supply current. Sometimes this arrangement will result in frequent tripping of the circuit breaker, in which case the tripping current may be reset to 15% of the rated current and/or a time delay device may be fitted to prevent tripping due to short duration transients. See {5.9} for more information concerning residual current devices,

2 the shell of the heater must be bonded to the sheath or armour of the supply cable with a conductor of current rating at least equal to the RCD tripping current, but with a minimum cross-sectional area of 2.5 mm².

C Single-phase electrode heater

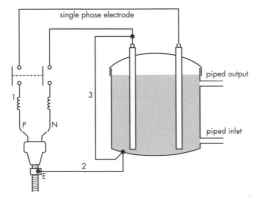

Fig 8.14 Single phase electrode heater

The requirements for single-phase electrode heaters are:
1 a double-pole linked circuit breaker with overload protection in each line,
2 the shell of the heater bonded to the sheath and/or armour of the supply cable with a conductor of current-carrying capacity at least equal to that of each live conductor,
3 the shell of the heater bonded to the neutral,
4 the supply must be one which has an earthed neutral.

D Insulated single-phase electrode heater, not permanently piped to water supply

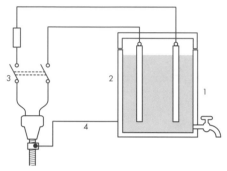

Fig 8.15 Insulated single-phase electrode heater, not piped to water supply

This small heater is usually fixed in position, but is filled by using a flexible hose or a water container. The requirements are:
1 there must be no contact with earthed metal,
2 the heater must be insulated and shielded to prevent the electrodes from being touched,
3 control must be by means of a single-pole fuse or by a circuit breaker and a double pole switch,
4 the shell of the heater must be bonded to the cable sheath.

8.12.3 Instantaneous water heaters [554.3]

Water heaters of this type are in general use to provide hot water for showers, making drinks, and so on. They transfer heat from the element directly to the water flowing over it, and therefore will be arranged to switch on only when water is flowing. In most cases, the element of the heater has a fixed rating, and so transfers energy in the form of heat to the water at a constant rate. The temperature rise of the water passing over the element therefore depends on the inlet water temperature and the rate of water flow. If the discharge rate is high, the energy provided by the element may be insufficient to raise the water temperature to the desired level.

The lower the rate of flow, the hotter will be the water at the outlet. This is the reason for the common complaint of being scalded whilst under the shower if someone turns on a tap elsewhere, reducing the water pressure and the rate of flow over the element. Some heaters are provided with an automatic cut-out to switch off the element if a preset outlet temperature is exceeded. {Figure 8.16} shows, in graphical form, the expected outlet water temperature for various water flow rates from 3 kW, 6 kW and 8 kW heaters assuming an inlet water temperature of 10°C.

Some heaters are provided with thyristors or triacs to continuously vary the heater rating to maintain a desired water outlet temperature so long as the variation in the rate of flow is not too great. These devices adjust the effective current flow by delaying the instant in each half-cycle of the supply at which current starts to flow.

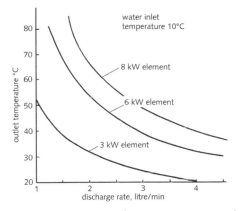

Fig 8.16 Outlet temperatures from instantaneous water heaters

The Regulations point out that a heater with an uninsulated element is unsuitable where a water softener of the salt regenerative type is used because the increased conductivity of the water is likely to lead to excessive earth leakage currents from the element. The agreement of the Water Supply Authority is usually needed before installation.

It is essential that all parts of this type of heater are solidly connected to the metal water supply pipe, which in turn is solidly earthed independently of the circuit protective conductor. A double pole linked switch must control the heater. In the case of a shower heater, if this switch is not built into the heater itself, a separate pull switch must be provided adjacent to the shower, with the switch itself being out of reach of a person using the shower. The arrangement of an instantaneous heater

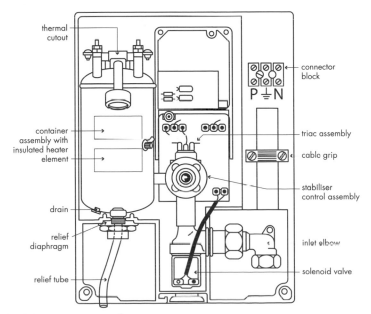

Fig 8.17 Instantaneous water heater

is shown in {Fig 8.17}. If the neutral supply to a heater with an uninsulated element is lost, current from the phase will return via the water and the earthed metal. Therefore, a careful check is necessary to ensure that there is no fuse, circuit breaker or non-linked switch in the neutral conductor. A problem frequently arises due to the increasing use of 3 kW instantaneous water heaters in caravans because supply systems have not usually been designed to allow for such heavy loading.

8.12.4 Surface, floor, soil and road warming installations [554.5]

Most cables have conductors of very low resistance so that the passage of current through them dissipates as little heat as possible. By using a cable with a higher conductor resistance (typical resistances for some of the resistive alloys used are from 0.013 to 12.3 Ω/m) heat will be produced and will transfer to the medium in which the cable is buried. There are many types of cable, a typical example being shown in {Fig 8.18}.

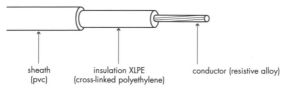

sheath   insulation XLPE   conductor (resistive alloy)
(pvc)   (cross-linked polyethylene)

Fig 8.18 One type of floor heating cable

Some of the many examples of the use of heating cables are:
1 space heating, using the concrete floor slab as the storage medium,
2 under-pitch application on sports grounds, to keep the playing surfaces free of frost and snow,
3 in roads, ramps, pavements and steps to prevent icing,
4 surface heating cables, tapes and mats, used for frost protection, anti-condensation heating, process heating to allow chemical reactions, drying, processing thermoplastic and thermosetting materials, heating transport containers, etc
5 in rainwater drainage gutters to prevent blocking by ice and snow, and
6 for soil warming to promote plant growth in horticulture.

The heating cables used in such situations must be able to withstand possible damage from shovels, wheelbarrows, etc. during installation, as well as the corrosion and dampness which is likely to occur during use. They must be completely embedded, and installed so that they are not likely to suffer damage from cracking or movement in the embedding material, which is often concrete. The loading of the installation must be such that the temperatures specified for various types of conductor {Table 8.7} are not exceeded. Where an electrical under-floor heating system is used in a bath or shower room, it must either have an earthed metallic sheath which is supplementary bonded or be covered by an overall earthed metallic grid which is similarly bonded.

Where heating cables pass through, or run close to, materials that present a fire hazard, they must be protected from mechanical damage by a fire-proof enclosure. Where normal circuit cables are run through a heated floor, they must have the appropriate ambient temperature correction factor applied {4.3.4}. Heating cables may be obtained ready jointed to normal cables at their ends (cold tails) for connection to the supply circuit as shown in {Fig 8.19}.

Table 8.7 Maximum conductor temperatures for floor warming cables

Type of cable	Maximum conductor operating temperature (°C)
General purpose pvc insulated	70
Enamelled conductor, pcp over enamel, pvc overall	70
Enamelled conductor, pvc overall	70
Enamelled conductor, pvc over enamel, lead alloy 'E' sheath overall	70
Heat resisting pvc over conductor	90
Nylon over conductor, heat resisting pvc overall	90
Synthetic rubber elastomeric insulation over conductor	90
Mineral insulation over conductor, copper sheath overall	*
Silicone-treated woven-glass sleeve over conductor	180

*Temperature depends on factors such as
1 type of seal used
2 if a cold lead-in section is used
3 the outer covering material if any, and
4 the material in contact with the heating cable.

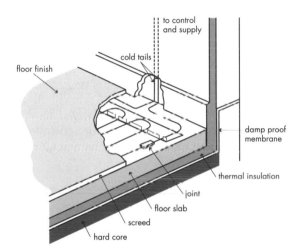

Fig 8.19 Floor heating installation

8.13 DISCHARGE LIGHTING

8.13.1 Low voltage discharge lighting [559]

The very high luminous efficiency of discharge lamps has led to their almost universal application for industrial and commercial premises; the introduction of low rated types as direct replacements for filament lamps has led to their wider use in domestic situations. Discharge lamps are those which produce light as a result of a discharge in a gas. Included are:

Fluorescent

Really low pressure mercury vapour lamps, very widely used for general lighting in homes, shops, offices, etc.

Compact fluorescent

These low power lamps are intended as direct replacements for filament lamps. Although first cost is higher, the increased efficiency (efficacy) and longer life makes them economically superior to filament types. The advent of Part L of the Building

Regulations, requiring more efficient lighting has made their use a legal requirement in many instances.

High pressure mercury
Provide a very intense lighting level for outside and internal industrial use in situations where the (sometimes) poor colour rendering is not important.

Low pressure sodium
The most efficient lamp of all, but its poor colour (orange) light output limits its use to street and external industrial lighting

High pressure sodium
The acceptable golden light colour enables the lamp to be used for road and outside lighting in areas where better colour rendering is needed, as well as for large indoor industrial applications.

Discharge lamps, unlike their incandescent counterparts, require control gear in the form of electronics, chokes, ballasts, auto-transformers and transformers. Electronic controls usually lead to deformation of the waveform and the introduction of harmonics, so where used in three-phase systems reduced neutral conductor cross-sectional areas are not permitted. These devices usually result in a lagging power factor, which is corrected, at least partially, by connecting capacitance across the supply. This control gear should be positioned as close as possible to the lamps. Because of low power factor and the inductive/capacitive nature of the load, switches should be capable of breaking twice the rated current of a discharge lamp system, and maximum demand is calculated by using a multiplying factor of 1.8 {6.2.1}.

Electronic devices are becoming increasingly common to provide high voltage pulses to assist discharge lamps to strike (start). These pulses can cause problems with insulation breakdown in some types of cable, particularly low voltage mineral insulated types.

8.13.2 High voltage discharge lighting
The Regulations have now omitted reference to these installations, which are usually neon signs or cold cathode lighting. Some sources suggest that the role assumed by this type of lighting (mainly for advertising purposes) is likely to be taken in future by extra-low voltage light emitting diodes (LEDs). Requirements for high voltage signs are to be found in BS 559. Any electrician who will be concerned with the high voltage aspects of such installations should therefore take expert advice, which is likely to include making reference to the BS.

Low voltage wiring to feed the transformers in this kind of installation is similar to most other installations except that special requirements apply to the Fireman's Switch which controls the installation and allows the fire service to make sure that high voltages have been isolated before playing water on the system. These special requirements have been covered in {3.2.2} and will not be repeated here.

8.14 UNDERGROUND AND OVERHEAD WIRING
8.14.1 Overhead wiring types [521]
Most overhead wiring is carried out by the Electricity Supply Companies, and is not in the domain of the electrician. However, wiring between buildings is often necessary, and can be carried out using one or more of the following methods:

1 elastomeric or thermoplastic insulated cables bound to a separate catenary wire,

2 elastomeric or thermoplastic insulated cables with an integral catenary wire,

3 thermoplastc insulated and sheathed, or elastomeric insulated with an oil-resisting and flame retardant sheath, provided that precautions are taken to prevent chafing of the insulation or undue strain on the conductors,

4 and 5 bare or thermoplastic insulated conductors on insulators. In this case, the height of the cables and their positions must ensure that basic protection (formerly called direct contact) is provided,

6 cable sheathed and insulated as in 3) above enclosed in a single un-jointed length of heavy-gauge steel conduit of at least 20 mm diameter. See {Table 8.8} for details.

Table 8.8 Maximum span and minimum height for overhead wiring between buildings (Based on Table G3 of Guidance Note 1)

Cable type (See {8.14.1})	Maximum span length (m)	At road crossings	Accessible to vehicles	Not accessible to vehicles*
1	3	—	—	3.5
2	3	—	—	3.5
3	3	—	—	3.5
4	30	—	—	3.5
5	No limit	—	—	3.5
6	3	5.8	5.2	3.0

*This column is not applicable in agricultural premises

8.14.2 Maximum span lengths and minimum heights [522]

The required information is contained in {Table 8.8} and {Fig 8.20}. Four special points are worth noting here.

1 in some locations where very tall traffic may be expected (e.g. yacht marinas, or where tall cranes are moved) heights above ground must be increased,

2 in agricultural situations, the lower heights of 3.5 m and 3.0 m are not permitted,

3 where caravan pitches are fed, overhead cables must be at least 2 m horizontally outside the bounds of each pitch,

4 overhead supplies to caravan pitches must have a minimum ground clearance of 6.0 m where there will be moving vehicles, and 3.5 m elsewhere.

NB All ground clearances are for positions not accessible to vehicular traffic.

8.14.3 Underground wiring [522]

Three types of cable may be installed underground:

1 armoured or metal sheathed or both,

2 p.v.c. insulated concentric type. Such a cable will have the neutral (and possibly the protective conductor) surrounding the phase conductor. An example is shown in {Fig 4.1(c)}.

3 Any suitable cable enclosed in conduit or duct which gives at least the same degree of mechanical protection as an armoured cable.

No specific requirements for depth of burial are given in the Regulations, except that the depth should be sufficient to prevent any disturbance of the ground reasonably likely to occur during normal use of the premises. Hence, a cable to outbuildings installed under a concrete path could be at 400 mm, whilst if running through a

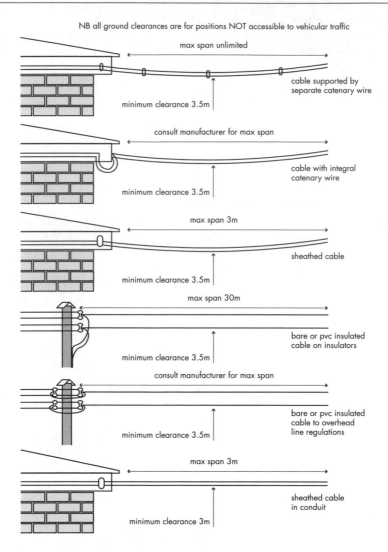

NB all ground clearances are for positions NOT accessible to vehicular traffic

max span unlimited

cable supported by separate catenary wire

minimum clearance 3.5m

consult manufacturer for max span

cable with integral catenary wire

minimum clearance 3.5m

max span 3m

sheathed cable

minimum clearance 3.5m

max span 30m

bare or pvc insulated cable on insulators

minimum clearance 3.5m

consult manufacturer for max span

bare or pvc insulated cable to overhead line regulations

minimum clearance 3.5m

max span 3m

sheathed cable in conduit

minimum clearance 3m

Fig 8.20 Overhead lines. Ground clearances shown are for places which are not accessible to vehicular traffic – see {table 8.8}.

cultivated space which could be subject to double digging would be less likely to disturbance if buried at 700 mm. For caravan pitches, cables should be installed outside the area of the pitch, unless suitably protected, to avoid damage by tent pegs or ground anchors.

Cables must be identified by suitable tape or markers above the cable, so that anyone digging will become aware of the presence of the cable. Cable covers may also offer both identification and protection as shown in {Fig 8.21}. It is often useful to lay a yellow cable marker tape just below ground level so that this will be exposed by digging before the cable is reached. Careful drawings should also be made to indicate the exact location of buried cables; such drawings will form part of the installation manual (see {7.2.1}).

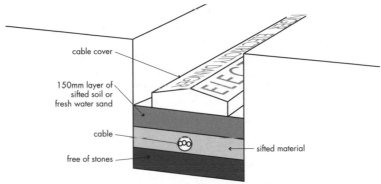

Fig 8.21 Cable covers

8.15 OUTDOOR INSTALLATIONS AND GARDEN BUILDINGS

8.15.1 Temporary and garden buildings [52]

Many dwelling houses have buildings associated with them that are not directly part of the main structure. These include garages, greenhouses, summerhouses, garden sheds, and so on. Many of them have an installation to provide for lighting and portable appliances. It is important to appreciate that the lightweight (and sometimes temporary) nature of such buildings does not reduce the required standards for the electrical installation. On the contrary, the standards of both the installation and its maintenance may need to be higher to allow for the arduous conditions.

Particular attention must be paid to the following:

1 supplies to such outbuildings must comply with the requirements for overhead and underground supplies stated in {8.14}, Garden layouts are very likely to change over a period of time, so all cables should be buried to a depth of at least 450 mm with a route marker tape at 150 mm. Cable runs must be recorded on careful drawings, and wherever possible should follow the edges of the garden plot,

2 the equipment selected and installed must be suitable for the environment in which it is situated. For example, a heater for use in a greenhouse will probably meet levels of humidity, temperature and spraying water not encountered indoors, and should be of a suitably protected type,

3 the earthing and bonding must be of the highest quality because of the increased danger in outdoor situations. All socket outlets must be protected by 30 mA RCDs.

8.15.2 Garden installations

Increasing use is being made of electrical supplies in the garden, for pond-pumping systems, lighting, power tools and so on. The following points apply:

1 Socket outlets installed indoors but intended to provide outdoor supplies must be protected by an RCD with a maximum operating current of 30 mA. Any portable equipment not fed from a socket outlet must also be protected by an RCD with a 30 mA operating current. Outdoor sockets also require the same RCD protection and must also satisfy IP44 requirements (see {2.4.3}).

2 Garden lighting, pond pumps and so on should preferably be of Class III construction, supplied from a SELV system and having a safety isolating transformer supply. Where 230 V equipment must be used, it should be Class II double insulated (no earth) and should be suitably protected against the

ingress of dust or water. If accessible Class I equipment is used its supply system must have an earth fault loop impedance low enough to allow disconnection within 0.4 s in the event of an earth fault.

3 Earthing must be given special attention. All buildings must be provided with 30 mA RCD protection, but the Electricity Supply Company should be consulted to ascertain their special requirements if the supply system uses the PME (TN-C-S) system. Where the supplier does not provide an earth terminal, each outbuilding must be provided with an effective adjacent earth electrode.

4 Outbuildings are often of light construction and therefore are subject to extremes as far as temperature swings are concerned. It is therefore important to bear this in mind when selecting equipment and components.

5 Extraneous conductive parts of an outbuilding which may become live due to a fault should be bonded to the incoming protective conductor.

6 Every outbuilding with an electrical supply should be provided with a means of isolation to disconnect all live conductors including the neutral.

7 All outbuildings where basic protection (formerly direct contact protection) is by earthed equipotential bonding and automatic disconnection of the supply should have a disconnection time in the event of an earth fault which does not exceed 0.4 s.

8 Cables which are not buried must be shielded from direct sunlight, whose ultra-violet content will affect plastics. Cables with ultra-violet protected sheaths may be used; the preferred colour for such cables is black.

9 Large garden ponds present a particular problem because of the probability that sooner or later people will fall into them. In such a case, the Regulations for swimming pools (see {8.3}) should be applied. Pumps and lighting should be SELV with the safety source at least 3.5 m outside the edge of the pond. All cables in the pond must be in ducts or conduits which are built into it and not be allowed to lie loose. All pond equipment must be protected to IP55 (see {2.4.3}) or better.

8.16 INSTALLATION OF MACHINES AND TRANSFORMERS
8.16.1 Rotating machines [552]

The vast majority of motors used in industry are of the three-phase squirrel-cage induction type. Smaller motors are usually single-phase induction machines. Induction motors have important advantages, such as robustness, minimal maintenance needs, and self-starting characteristics, but all draw very high starting currents from their supplies {Fig 8.22}. This starting current is a short-lived transient, and may usually be ignored when calculating cable sizes.

Although the starting current may be several times the running current, the value depending on the machine characteristics and the connected mechanical load, its short duration will not lead to overheating in usual circumstances. If frequent starting is a requirement, larger supply cables may be necessary to avoid damage to insulation. A problem could arise when fast-acting fuses or circuit breakers are used for short-circuit protection; the high starting current may result in operation of the protective device. A common, but unsatisfactory, remedy for this difficulty is to increase the rating of the protective device, leading to a loss of proper overload protection. A possible solution is to use dual rated fuses (gM types). For example, a 25M40 fuse has a continuous rating of 25 A and the operating characteristics of a 40 A fuse.

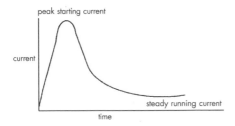

Fig 8.22 Starting current of an induction motor

A word is necessary concerning motor ratings. Many years ago it was decided to replace the horsepower as the unit of output power with the kilowatt. Unfortunately, the old horsepower is a very long time in dying. Many machines still have rating plates giving output power in horsepower. The conversion is straightforward. Since one horsepower is the same as 746 W, horsepower is converted to kilowatts by multiplying by 0.746.

It is sometimes practice to stop a motor very quickly by feeding it with a reverse current. When this method is provided it is important that the machine does not begin to move in the reverse direction if this would cause danger.

Where other types of motor, such as wound rotor and commutator induction or thyristor fed dc types are used, the cables must be suitable for carrying running currents on full load, which will usually mean that they are large enough to carry the short duration starting currents.

Every motor rated at 0.37 kW (0.5 horsepower) or more must be fed from a starter that includes overload protection. Such devices have time-delay features so that they will not trip as a result of high starting current, but will do so in the event of a small but prolonged overload. They have the advantages over fuses and single-pole circuit breakers that all three lines of a three-phase system are tripped by an overload in any one of them. If only one line were broken, the resulting 'single-phasing' operation of the motor could cause it to overheat.

It is often necessary to provide a means to prevent automatic restarting after failure of the supply. For example, if the supply to a machine shop fails, the machine operators are likely to use the enforced break in production to clean and service their machines. If so, when the supply is restored, the presence of hands, brushes, tools, etc. in the machines when they automatically restart would cause serious danger. The necessary 'no-volt protection' is obtained by using a starter of the type

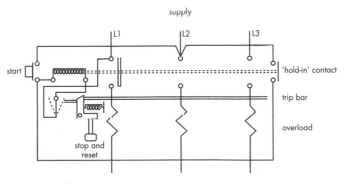

Fig 8.23 Direct-on-line starter

whose circuit is shown in {Fig 8.23}. The coil is fed through the 'hold-in' contacts, which open when the supply fails; only pressing the 'start' button can then operate the motor. This requirement does not apply to protected motors that are required to restart automatically after mains failure. Examples are motors supplying refrigeration and pumping plants. It is important that lock-off stop buttons are not used as a means of isolation.

8.16.2 Transformers [555]

Transformers are used more widely in the supply system than in the installation itself except for very large installations. The link between the primary and secondary windings of a double wound transformer is magnetic, not electrical, so there is no electrical connection with the supply system, or with its earthing system, from a circuit fed by the secondary winding of a transformer. This loss of earthing may be an advantage where electrical separation is required, for example, where an electric shaver in a bathroom is fed from the secondary winding of a transformer (see {5.8.4}).

Step-up transformers must not be used in IT systems. In systems where they are permitted, linked multipole switches must be provided so that the supply is simultaneously disconnected from all live conductors, including the neutral.

8.17 REDUCED VOLTAGE SYSTEMS

8.17.1 Types of reduced voltage [414]

Most installations operate at low voltage, which is defined as up to 1000 V ac or 1500 V dc between conductors, and up to 600 V ac or 900 V dc between conductors and earth. Extra-low voltage is defined as not exceeding 50 V ac or 120 V dc between conductors or from conductors to earth. Five types of reduced voltage system, all intended to improve safety, are recognised by the Regulations. They are:

  1 separated extra-low voltage or SELV {8.17.2},
  2 protective extra-low voltage or PELV {8.17.3},
  3 functional extra-low voltage or FELV {8.17.4},
  4 reduced voltage {8.17.5},
  5 voltages at values lower than the maximum for ELV may be required to ensure the safety of people and of livestock in areas where body resistance is likely to be very low (see {8.7}).

8.17.2 Separated extra-low voltage (SELV) [414.4]

The safety of this system stems from its low voltage level, which should never exceed 50 V ac or 120 V dc, and is too low to cause enough current to flow to provide a lethal electric shock to people unless their body resistance is much reduced. The reason for the difference between ac and dc levels is shown in {Figs 3.9 & 3.10}.

It is not intended that people should make contact with conductors at this voltage; where live parts are not insulated or otherwise protected, they must be fed at the lower voltage level of 25 V ac or 60 V ripple-free dc although insulation may sometimes be necessary, for example to prevent short-circuits on high power batteries. Sources of SELV (and PELV) can be

  1 a safety isolating transformer complying with BS EN 61558-2-6, or
  2 a motor generator with windings that give the same protection as in 1, or
  3 an electro-chemical battery, or
  4 electronic devices that are incapable of providing higher voltages than those

specified above. An exception is devices (such as the insulation resistance tester) whose output voltage falls to a safe level as soon as it delivers a significant current.

To qualify as a separated extra-low voltage (SELV) system, an installation must comply with conditions, which include:

1 it must be impossible for the extra-low voltage source to come into contact with a low voltage system, using basic insulation or protective separation from live circuits using double or reinforced insulation.

2 there must be no connection whatever between the live parts of the SELV system and earth or the protective system of low voltage circuits. The danger here is that the earthed metalwork of another system may rise to a high potential under fault conditions and be imported into the SELV system.

3 there must be physical separation from the conductors of other systems, the segregation being the same as that required for circuits of different types (see {6.6}).

4 plugs and sockets must not be interchangeable with those of other systems; this requirement will prevent a SELV device being accidentally connected to a low voltage system,

5 plugs and sockets must not have a protective connection (earth pin). This will prevent the mixing of SELV and FELV devices. Where the Electricity at Work Regulations 1989 apply, sockets must have an earth connection, so in this case appliances must be double insulated to class II so that they are fed by a two-core connection and no earth is required,

6 luminaire support couplers with earthing provision must not be used. Additional protection may be provided by the use of 30 mA RCDs, (although they must not be the sole means of protection) and by supplementary equipotential bonding.

8.17.3 Protective extra-low voltage (PELV) [414.4]
The protective extra-low voltage system (PELV) differs from the separated extra-low voltage system (SELV) system only in that it is connected to earth. This connection may be achieved by connection to an earthed protective conductor within the system itself, or by connection to the main earthing terminal.

8.17.4 Functional extra-low voltage (FELV) [411.7]
Where extra-low voltage is used but all the requirements for SELV or PELV systems are not met or are not necessary, then functional extra-low voltage (FELV) may be applied.

If a separated extra-low voltage (SELV) system is earthed and if the insulation of the supply that feeds it does not meet the necessary requirements, it ceases to be a SELV system and becomes a functional extra-low voltage (FELV) system. {Fig 8.24} illustrates this difference.

Such metalwork must, however, be bonded and earthed if the insulation between the low-voltage supply circuit and the functional extra low voltage circuit does not meet the requirements stated above. Plugs and sockets in FELV systems must not be interchangeable with those of other supply systems in use in the same premises.

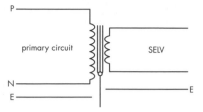

becomes a functional extra-low voltage system when:

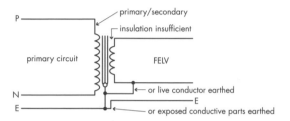

Fig 8.24 Earthing relationship, SELV and FELV systems

8.17.5 Reduced voltage [411.8]

In situations where the power requirements are high, extra-low voltage systems would need to deliver very high currents. If an increased voltage can still lead to a safe system, the current required will be reduced. We are therefore considering a voltage higher than extra-low but lower than low voltage.

Table 8.9 Maximum earth-fault loop impedance (Ω) for maximum discon-nection time of 5 s in reduced voltage systems
(From Table 41.6 of BS 7671:2008)

| | Circuit breakers to BS EN 60898 and RCBOs to BS EN 61009 | | | | | | General purpose fuses BS 88 | |
| | Type B | | Type C | | Type D | | | |
Uo (V)	55	63.5	55	63.5	55	63.5	55	63.5
Rating (A)								
6	1.83	2.12	0.92	1.07	0.47	0.53	3.20	3.70
10	1.10	1.27	0.55	0.64	0.28	0.32	1.77	2.05
16	0.69	0.79	0.34	0.40	0.18	0.20	1.00	1.15
20	0.55	0.64	0.28	0.32	0.14	0.16	0.69	0.80
25	0.44	0.51	0.22	0.26	0.11	0.13	0.55	0.63
32	0.34	0.40	0.17	0.20	0.09	0.10	0.44	0.51
40	0.28	0.32	0.14	0.16	0.07	0.08	0.32	0.37
50	0.22	0.25	0.11	0.13	0.06	0.06	0.25	0.29
63	0.17	0.20	0.09	0.10	0.04	0.05	0.20	0.23

The highest voltage permitted for such systems is 110 V between conductors for both single- and three-phase systems, with voltages to earth of 55 V for single-phase and 63.5 V for three-phase supplies {Fig 8.25}.

The reduced voltage supply can be taken from a double wound isolating transformer, or from a suitable motor generator set, provided it has the unusual feature of a centre-tapped winding. The system must be insulated and protected against basic contact as for a low voltage installation. Earth-fault loop impedance must allow automatic disconnection in a maximum of 5 s (see {Table 8.9}), or an RCD with an operating current of no more than 30 mA must be provided. The result of multiplying earth-fault loop impedance (Ω) by the RCD operating current (A) must not exceed 50 (V). Plugs, sockets and cable couplers in the reduced voltage system must all be provided with protective conductor contacts (earth pins) and must not be interchangeable with those of any other system in use at the same location.

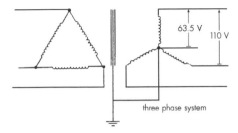

63.5 V 110 V

three phase system

Fig 8.25 Reduced voltage systems

8.18 MARINAS

8.18.1 Introduction [709]

These locations have been included in BS 7671 for the first time in the 17th Edition. A marina is a location, often a harbour, for leisure craft to berth. Like residential caravans, such craft require external power supplies, and this section is intended to ensure the safety and standardisation of such supplies. The marina will often include shore-based facilities such as offices, workshops, toilets, leisure accommodation and so on; such buildings are required to comply with the general requirements of the Regulations.

The electrical installation of a marina is subject to hazards not usually encountered elsewhere, such as the presence of water and salt, the movement of the craft, increased corrosion due to the presence of salt water and dissimilar metals, and the possibility of equipment being submerged due to unusual wave activity in bad weather. Electrical risks are also greater when craft are supplied from a marina, including

1   open circuit faults on the PEN conductor of PME supplies raising the potential to true earth of all metalwork (including the craft) to dangerous levels.

2   lack of an established equipotential zone external to the craft.

3   loss of earthing due to long supply cable runs and flexible cord connectors exposed to the weather and to mechanical damage.

Such dangers are reduced by the use of 30 mA RCDs in both the craft and the marina installations.

8.18.2 The marina electrical installation [709]

As for caravans, the neutral of a PME system must not be connected to the earthed system of a boat so that the hazards that follow the loss of continuity in the supply PEN conductor are avoided. This rules out the use of PME supplies for marinas. Where this is the supply provided, it must be converted to a TT system at the main distribution board by provision of a separate earth electrode system of driven rods or buried mats with no overlap of resistance area with any earth associated with the PME supply. If the marina is large enough, it may be that the supply company will provide a separate transformer and a TN-S system.

Isolating transformers may be used to supply craft, having the advantage of reducing electrolytic corrosion. The transformer must comply with BS EN 60742 or with BS EN 61558. The protective conductor of the shore supply must not be connected to the bonding of the craft, which must instead be connected to one of the secondary winding terminals of the isolating transformer.

Shock protection
Protection by obstacles or by placing out of reach are not acceptable as methods of preventing electric shock under normal conditions (previously known as direct contact): such protection must be provided by the insulation of live parts and/or by the use of barriers and enclosures. An RCD must be used to reduce the risk of electric shock caused under normal conditions, but must not be used as the sole means of protection. A non-conducting location must not be used for protection from fault contact (previously known as indirect contact).

Wiring systems
No aluminium cables may be used in marina installations. Acceptable cables for fixed wiring are:
- those with thermoplastic, thermosetting or elastomeric insulation and sheath installed in non- flexible non-metallic conduit, or in heavy-duty galvanised conduit,
- mineral insulated cable with extruded thermoplastic (p.v.c.) covering,
- armoured cables with sheaths of thermoplastic or elastomeric material,
- underground cables, buried at a depth of at least 0.5 m,
- overhead insulated cables at a height of 6 m where vehicle traffic is expected, or 3.5 m otherwise,
- armoured cables with a serving of elastomeric or thermoplastic material.

For wiring on or above a jetty, wharf, pier or pontoon, overhead cables with a catenary, conduit or trunking systems and mineral insulated wiring must **not** be used.

A boat will move relative to the land due to wind and waves. Cables must be selected and installed so that mechanical damage as a result of tidal and other movement of floating structures is prevented. Cabling to bridge ramps, pontoons and movable jetties must be carried out in flexible cables with EPR or thermosetting insulation and sheaths. Overhead lines are not permitted.

It should be noted that p.v.c. cables are not usually suitable for continuous immersion in water. Cables intended to be permanently immersed to a depth of 4 m or more should be sheathed with lead; where the immersion is less than 4 m, cables should be armoured and provided with polythene bedding and outer sheath. Due to possible problems with corrosion, cable armouring must not be relied upon as a circuit protective conductor. A separate CPC must be provided to which the armouring is securely connected. Unless the buried cables are above the water table they must be suitable for continuous immersion.

Distribution boards
Distribution boards must be protected to IP44, and against the ingress of dust and sand. Construction of glass reinforced plastic (GRP) gives better corrosion protection than galvanised steel. Where mounted on fixed or floating jetties, they must be at a minimum of 1 m above the high water level although in the latter case this may be reduced to 300 mm if protection from splashing is provided which does not provide a tripping hazard. Boards must be fitted with locks to prevent unauthorised

access, and with internal barriers to prevent contact with live parts (to IP 2X) when the doors are open. Low power heaters may be needed within the boards to prevent excessive condensation.

Socket outlets

A common installation method is to provide a feed from the shore to a floating pontoon via a bridge or ramp, and then to equip the pontoon with socket outlets to feed the craft moored to it. Socket outlets may be single- or three-phase. Where multiple single-phase sockets are installed on the same pontoon, they must all be connected to the same phase of the supply unless fed through isolating transformers. All socket outlets must be to BS EN 60309-2 with keyways at 6h, protected to IPX4 minimum, coloured red if three-phase and blue if single-phase. Single-phase sockets are usually rated at 16 A and three-phase at 32 A, although higher ratings may be installed where the need arises. Socket outlets should be positioned as close as possible to the berth of the vessel they feed, with a minimum of one socket per berth, although up to four sockets may be provided in a single enclosure. Each socket outlet must be provided with a means of isolation that breaks all poles on TT systems, and must be protected by an overcurrent device such as a fuse or a circuit breaker. Every socket must be individually protected by a 30 mA RCD. Each socket or group of sockets must be provided with a durable and legible notice giving instructions for the electricity supply (see {Table 8.10}); alternatively the notice must be placed in a prominent position or must be issued to each berth holder.

Inspection and testing

Inspection and testing must be carried out in full accordance with BS 7671 (see {Chapter 7}). Periodic testing must be at least annually, but may need to be more frequent if the marina is exposed or subject to misuse.

8.19 MEDICAL LOCATIONS

Medical locations do not appear in the 17th Edition of BS 7671, so this section is based on Chapter 10 of Guidance Noted 7 published in 2001.

8.19.1 Introduction

Medical locations are often hospitals, but may also include private clinics, medical and dental surgeries, health-care centres and dedicated medical areas in the workplace.

The use of medical electrical equipment is split into three categories:

1 Life-support: infusion pumps, dialysis machines and ventilators;
2 Diagnostic: X-ray machines, magnetic resonance imagers, blood pressure monitors, electroencephalograph (EEG) and electrocardiograph (ECG) equipment;
3 Treatment: surgical diathermy and defibrillators.

This equipment is increasingly used on patients under acute care. Supplies to it require enhanced reliability and safety, as well as measures to reduce incidents of electric shock.

Shock hazards are covered in {3.4} and indicate that currents as low as 10 mA passing through the human body can result in muscular paralysis (an inability to move) followed by respiratory paralysis (an inability to breathe). At about 20 mA ventricular fibrillation (loss of normal heart muscle rhythm) may occur and may be fatal.

Table 8.10 Instructions for electricity supply

Berthing instructions for connection to the shore supply
   This marina provides power for use on your leisure craft with a direct connection to the shore supply which is connected to earth. Unless you have an isolating transformer fitted on board to isolate the electrical system of your craft from the shore supply system, corrosion through electrolysis could damage your craft or surrounding craft.

ON ARRIVAL
1 Ensure that the supply is switched off before inserting the craft plug.
2 The supply to the berth is *** V, ** Hz. The socket outlet will accommodate a standard marine plug coloured ****
3 For safety reasons, your craft must not be connected to any other socket outlet than that allocated to you and the internal wiring on your craft must comply with the appropriate standards.
4 Every effort must be made to prevent the connecting flexible three-core cable from falling into the water if it should become disengaged. For this purpose, securing hooks are provided alongside socket outlets for anchorage at a loop of tie cord.
5 For safety reasons, only one leisure-craft connecting cable may be connected to any one socket outlet.
6 The connecting flexible cable must be in one length, without signs of damage, and not containing joints or other means to increase its length.
7 The entry of moisture and salt into the leisure-craft inlet socket may cause a hazard. Examine carefully and clean the plug and socket before connecting the supply.
8 It is dangerous to attempt repairs or alterations. If any difficulty arises, contact the marina management.

BEFORE LEAVING
1 Ensure that the supply is switched off before the connecting cable is disconnected and any tie cord loops are unhooked.
2 The connecting flexible cable should be disconnected first from any marina socket outlet and then from the leisure craft inlet socket. Any cover that may be provided to protect the inlet from weather should be securely replaced. The connecting flexible cable should be coiled up and stored in a dry location where it will not be damaged.

**** Appropriate figures and colours must be inserted
     either 220 – 250 V blue,
     or 380 – 415 V red.

The impedance of the human body (see {3.4.2}) may be considerably reduced during certain clinical procedures. For example, the skin resistance is reduced during surgery and defensive capacity either reduced by medication or removed altogether while anaesthetised. Special hazards occur during heart surgery, where electrical conductors may be placed within the heart , or with the use of conductive catheters. Under these conditions currents as low as a few tens of micro-amperes may well result in death.

The risk of electric shock is not the only danger in these situations. Often the loss of supply to such vital equipment as life support systems can be equally catastrophic, and such medical locations require very secure supplies. In some cases., conventional protective systems, such as RCDs, may necessarily have to be restricted.

8.19.2 Recommendations for medical locations
Patient safety
Since the hazard to people will depend on the treatment being administered, hospital locations are divided into groups as follows:

1 Group Zero: where no treatment or diagnosis using medical electrical equipment is administered, e.g. consulting rooms;
2 Group 1: where the electrical parts must necessarily make contact with the patient externally, or internally except where Group 2 applications are intended;
3 Group 2: where electrical parts are used in applications such as intracardiac procedures, in operating theatres or vital treatment where failure of the supply may cause danger to life.

TN-C systems (where earth and neutral are combined – see {5.2.5}) is not allowed in medical locations because load currents in PEN conductors and parallel paths can cause electromagnetic interference to medical equipment. Basic protection (formerly known as direct contact) must be by insulation of live parts and by barriers or enclosures. SELV and PELV systems (see {3.4.4, 8.17.2, 8.17.3}) may be used in Group 1 and in Group 2 locations, but are limited to 25 V r.m.s. a.c. or 60 V ripple free d.c. Even at these voltages, protection of live parts by insulation or by barriers or enclosures is essential. Exposed conductive parts of PELV systems must be connected to the local equipotential bonding conductor. FELV systems (see {3.4.4, 8.17.4}) may be used in Group Zero locations only.

Fault protection (previously called indirect contact) may be by automatic disconnection of the supply (see {3.4.6}), by electrical separation (see {5.8.4}), or by the use of Class II equipment (see {5.8.1}) except as indicated below:

For TN systems (see {5.2.3} and {5.2.4}) any medical equipment in Group 1 or in Group 2 locations which is situated within 2.5 m above the floor or within 1.5 m horizontally from the edge of the bed, operating table, etc., on which the patient is lying, called the patient environment (see {Fig 8.26}), must be RCD protected, the rating of the device being 30 mA or less if the circuit protection does not exceed 32 A, or 300 mA or less if the circuit protection exceeds 32 A. Note, however, that in Group 2 locations, RCDs may only be used for X-ray units or for equipment with a rated power greater than 5 kVA.

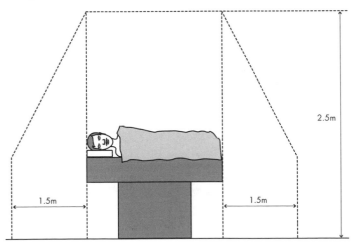

Fig 8.26 The patient environment

For TT systems (see {5.2.2} used in Group 1 and Group 2 situations, the same requirements apply as in TN systems with earth fault protection provided by RCDs.

For IT systems (see {5.2.6} it is recommended that such a system is used for circuits supplying medical equipment that is intended to be used for life-support of patients. This is because the total absence of an earthing system makes indirect contact an impossibility. Any medical equipment in Group 2 which is situated in the patient environment (see {Fig 8.26}) must have a safety isolating transformer incorporated into its IT system, and must have an insulation monitoring device. This device must have:

1 acoustic and visual alarms that are triggered at the first earth fault, and
2 an a.c. internal resistance of at least 100 kΩ, and
3 a test voltage not exceeding 25 V d.c., and
4 a test current not exceeding 1 mA even under fault conditions, and
5 activation when the insulation resistance falls to less than 50 kΩ, and
6 activation when any wiring or earth is disconnected.

If a socket outlet on a patient bedhead location in a Group 2 situation is supplied from an IT system, it must be fed from at least two separate circuits, or must be individually protected against overcurrent. Socket outlets in a Group 2 situation fed from other systems (TN-S or TT) must be clearly marked to distinguish them from IT system socket outlets. Any wiring system within a Group 2 location must supply only the equipment and fittings in that group (see {Fig 8.27}. Sockets and switches must be at least 0.2 m horizontally from medical gas outlets.

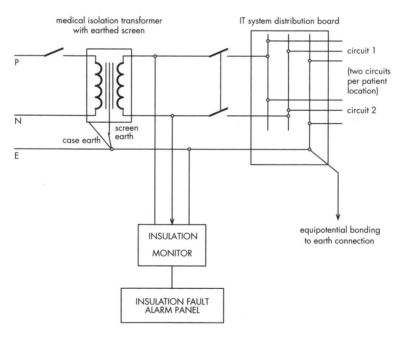

Fig 8.27 Distribution network with insulation monitoring

Supplementary equipotential bonding

This bonding is required within Groups 1 and 2 locations, the resistance of the bonding conductors not exceeding 0.2 Ω. The equipotential bonding busbar must be in or near the medical location, with connections clearly visible and readily disconnected.

Explosion risk

The gases used as anaesthetics in operating theatres are flammable if present in high concentrations. Provided that there is adequate ventilation (20 air changes per hour) no special precautions are necessary for the electrical installation.

Standby power supplies

The failure of the power supply may well have fatal consequences, for example if the lighting fails in an operating theatre, or the feed to a life support system is lost in an intensive care unit. Safety power supplies are split into five categories depending on how quickly the supply is restored after failure. The time periods are zero, for medical and microprocessor equipment, 0.15 s or 0.5 s for other medical equipment and theatre lighting, 15 s where a standby generator will take over supplies, or more than 15 s, which is likely to apply to offices, staff accommodation, and so on. Emergency lighting fed from a safety source is required at escape routes, exit signs and locations of essential services in Groups 1 and 2 locations. The power source must be capable of maintaining the supply to luminaires over and around operating theatre tables for three hours.

Department of Health recommendations

The Department provides guidance through its Health Technical Memoranda (HTM) series of publications. These extremely detailed volumes are available from The Stationery Office Ltd (TSO), formerly HMSO.

8.20 TEMPORARY ELECTRICAL INSTALLATIONS FOR STRUCTURES, AMUSEMENT DEVICES AND BOOTHS AT FAIRGROUNDS, AMUSEMENT PARKS AND CIRCUSES

8.20.1 Introduction [711, 740]

These Regulations deal with temporary electrical installations in exhibitions shows and stands, both indoor and outdoor within temporary or permanent structures. They have been included in BS 7671 for the first time with the publication of the 17th Edition. There will be increased risks in these situations due to:

1 the temporary nature of the installations,
2 severe mechanical stresses,
3 lack of permanent structures,
4 access by the general public, and
5 the fact that they are frequently dismantled and re-erected.

Since these risks are greater than those usually encountered, additional measures are necessary.

8.20.2 Recommendations for exhibitions, shows and stands [711, 740]
Shock protection

Basic protection (previously known as direct contact protection) (see {3.4.5}) by means of obstacles and placing out of reach must not be used except for 'dodgems', and neither must non-conducting location and earth-free equipotential bonding be used as fault protection (previously referred to as indirect contact, see {3.4.6}). Automatic disconnection of the supply is the preferred method, but because of the difficulty of bonding all accessible extraneous conductive parts, PME supplies (TN-C-S system) and protective earth and neutral (PEN) systems {5.7.2} must not be used for temporary or outdoor installations. A TN-S system (where the supply company provides an earth) could be used, but a TT system (earthed using an electrode) is preferable. Livestock are often present at agricultural shows. In this case, supplementary bonding must connect all exposed conductive parts and extraneous conductive parts that can be touched by them.

Distribution circuits which are at an increased risk of damage must be protected by an RCD with a 300 mA rating; to provide discrimination with other RCDs protecting final circuits, they should be time delayed. All final circuits for socket outlets rated at up to 32 A and all circuits other than for emergency lighting, must be protected by RCDs rated at 30 mA. However, loss of light in a crowded space is likely to cause a serious hazard, and such areas should be fed from at least two separate circuits with separate RCD protection.

Fire protection

Because of the increased risk of fire and burns in these locations, it is very important to follow the advice given in {3.5}. Stored material, such as cardboard boxes and fodder, are a special fire hazard. Remotely or automatically controlled motors must be fitted with manual reset controls against excess temperature. Lighting equipment, such as incandescent lamps, spotlights, projectors, etc., must be suitably positioned and guarded to prevent overheating to themselves or of adjacent surfaces. Sufficient ventilation must be provided to prevent the build-up of heat. See also {3.9}.

Isolation

Every separate temporary structure and each distribution circuit supplying outdoor installations must be provided with its own easily accessible and clearly identified means of isolation that must break phase and neutral conductors.

Equipment

Equipment must be mounted away from positions that may not be weatherproof. Tent poles are the most obvious place to mount equipment, but are often the weak point in the weather tightness of temporary structures. Switchgear should be mounted in closed cabinets that can only be opened by a key or a tool, but isolators must be always accessible.

Wiring systems

The types and protection of cables is of particular importance, as is the current-carrying capacity. Wherever there is a risk of damage, mechanical protection or armoured cables should be used. Underground cables should be clearly marked and protected against mechanical damage. Where luminaires are mounted below 2.5 m (within arm's reach) they must be firmly fixed and carefully sited to reduce danger to people and the possibility of fire. Electric motors must have isolation equipment breaking all poles, and must be provided with adjacent emergency stop systems. Socket outlets are subject to particular abuse by users in these situations. They should:

1 be adequate in number,
2 preferably not be floor mounted, but if so, must be mechanically protected and waterproof,
3 have no more than one flexible cable or cord connected to each plug.

Safety systems

Where a generator is used to supply the temporary installation, there must be full earthing using separate earth electrodes; they must be located or protected so as to prevent danger from hot surfaces or moving parts. For TN systems, all exposed conductive parts must be bonded back to the generator. Where the exhibition is held in a building, care must be taken to ensure that temporary structures do not impede escape routes. Where an exhibition is constructed out-of-doors, a fire alarm system must be installed to enclosed areas, and emergency lighting must be provided for escape routes.

Inspection and testing

The electrical installations of all stands must be re-tested on site after each assembly. Exhibitors and stallholders must be encouraged to visually check electrical equipment for damage on a daily basis.

8.21 Small-scale Embedded Generators (SSEG)

NOTE. This type of power supply system is novel, and anyone concerned with the installation of such a system would be well advised to seek the advice of both its manufacturer and of the Supply Company. Section [551] applies, and the following is based on it and on Chapter 18 of Guidance Noted 7.

8.21.1 Introduction [551, 552]

There is an increasing number of embedded generator systems in electrical installations. The term 'small scale' is taken to mean systems with a maximum current output of 16 A single phase (230 V), or 16 A per phase for a three-phase (400 V) system. Engineering Recommendation G83 by the Electricity Association has also provided guidance, which is followed here. G83 incorporates forms defining the information needed by the Supply Company, acceptance of which means that the installation will satisfy the Electricity Safety, Quality and Continuity Regulations 2002 (ESQC).

The likely types of embedded generators include

– Domestic combined heat and power (CHP) devices, usually gas boilers combined with Stirling engines
– Hydro systems, where the power is derived from water using a water-wheel or a turbine
– Wind power, the derivation being from a windmill
– Fuel cells, the fuel being hydrogen gas
– Photovoltaic systems, with energy from the sun (see {8.22}).

Fuel cells are expensive at present, but the prices are falling and a number of installations are in use; small numbers of hydro and wind powered systems are already in operation. Photovoltaic power supply systems are the subject of section {8.22} of this Chapter.

The form of embedded generator already becoming common is the type powered by a Stirling engine incorporated into a domestic gas boiler. This system uses heat from the gas to drive the engine, which in turn drives an electrical generator. The requirements of BS 7671 [551] (Low Voltage Generating Sets) must also be followed.

The systems considered here must meet the requirements of the ESQC Regulations as they are embedded generators. So long as the output of such a system does not exceed 16 A per phase they are classed as small scale embedded generators (SSEG) and are exempt from certain of the requirements provided that

a) The equipment is properly type tested and approved
b) The consumer's installation complies with BS 7671
c) The equipment disconnects itself from the supply network in the event of a fault, and
d) The supply company is advised of the generator to be used in parallel with their system.

8.21.2 Installation requirements [551]

All wiring must, of course, comply with BS 7671. A suitably rated overcurrent fuse or circuit breaker must protect the wiring from the electricity supply terminals to the SSEG, which must also be provided with an accessible local isolating switch, manually operated and capable of being secured in the 'off' position. (See {Fig 8.28})

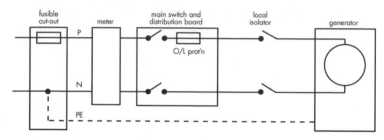

Fig 8.28 Isolation of the small scale embedded generator

When operating in parallel with a distributor's network, the generator windings must not be connected to earth. The presence of the embedded generator must be clearly indicated by labels positioned at

a) the incoming supply terminals
b) the meter position
c) the consumer unit, and
d) at all points of isolation.

The label used must be 120 mm x 35 mm, and worded as in Fig 8.29.

In addition to these labels, there must be a circuit diagram such as that in {Fig 8.28}, together with the settings of the protection used and details of the owner and maintainer of the generator displayed at the incoming supply position. It is an important responsibility of the user that this information is kept up to date. Operating instructions must contain the manufacturer's contact details.

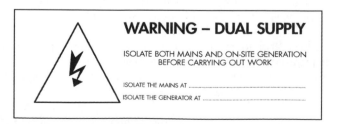

Fig 8.29 Dual supply notices for the SSEG

8.22 PHOTOVOLTAIC POWER SUPPLY SYSTEMS

NOTE This type of power supply system is novel, and anyone concerned with the installation of such a system would be well advised to seek the advice of both its manufacturer and of the Supply company.

8.22.1 Introduction [551, 712]

As indicated in the previous section, photovoltaic power supply systems are becoming widespread. The PV systems covered by BS 7671 must meet the requirements of the Electricity Safety, Quality and Continuity Regulations 2002 (ESQCR) as they are embedded generators. (See also {Section 8.21} concerning other embedded generators). So long as the output of such a system does not exceed 16 A per phase they are classed as small scale embedded generators (SSEG) and are exempt from certain of the requirements of ESQCR provided that

1   The equipment is properly type tested and approved
2   The consumer's installation complies with BS 7671
3   The equipment disconnects itself from the supply network in the event of a fault, and
4   The supply company is advised of the installation.

It follows from the above that we are usually concerned here with relatively small systems, such as those used in domestic or in small commercial installations.

8.22.2 Principles

It is not the purpose of this work to describe the operation of the photovoltaic cell, and no attempt will be made to do so. However, it will be appreciated that PV systems provide their output in the form of extra-low voltage DC, and electronic inversion systems will be necessary to change the output to alternating current at a frequency and voltage (AC) that will

1   make it compatible with the consumer's equipment, and
2   allow the PV supply to be operated in parallel with the supply system.

It will thus be apparent that protection for the user and for the installation will be required on both the AC and the DC sides of the system. The DC voltage of an individual PV cell is low, but cells may be series connected in 'strings', providing much higher, and sometimes dangerous, voltages.

8.22.3 Protection systems [712]

It is very important to appreciate that the PV system is a generator in its own right, and the output can still be dangerous even when the mains supply is removed. For SELV and PELV, rules against basic and fault protection (formerly called direct and indirect contact) apply and the direct voltage must not exceed 120 V. Fault protection on the DC side requires Class II or equivalent insulation on the DC side. Where there is no separation between the AC and the DC sides, RCD protection by a Type B device (see {5.9.2}) must be provided, but this is unnecessary where the PV inverter is unable to feed direct current into the AC installation. Fault protection by non-conducting location or by earth-free equipotential bonding is not permitted on the DC side. Overload protection on the DC side may be omitted when the current carrying capacity of the cable is 25% greater than the possible short circuit current it could carry; this requirement only applies to protection of cables. The manufacturer should be consulted concerning protection of PV modules.

Electromagnetic interference (EMI) can be a problem, but can be reduced by keeping the area of all wiring loops as small as possible. PV modules will be connected in series to provide a higher overall voltage, which should not exceed 120 V. Where blocking diodes are used, they should be rated at twice the maximum voltage of the string concerned. To facilitate maintenance, a means of isolating the PV inverter from both the AC and the DC sides must be provided. All junction boxes carrying supplies from the PV generator must carry warning notices to draw attention to the fact that live parts may still be present after disconnection from the inverter. Where equipotential bonding conductors are installed, they should be parallel to and as close as possible to live cables. Earthing of one of the live conductors on the DC side is permitted provided that there is separation between the AC and the DC sides.

8.23 FLOOR AND CEILING HEATING SYSTEMS

8.23.1 Introduction

These requirements of BS 7671 apply to both direct acting and thermal storage systems. They do not apply to wall heating systems, or to those for use outdoors (but see {8.12.4}).

8.23.2 Electric shock and overheating protection [753]

These systems are subject to risks not found in other installations, such as the penetration of nails, drawing pins etc. into a ceiling, or of carpet gripper nails etc. into a floor. For these reasons, all such installations must have basic protection (which was previously called direct contact protection) by the use of 30 mA RCDs. A problem may arise due to unwanted tripping of RCDs due to leakage current; this can often be overcome by limiting the rated heating power protected by a 30 mA RCD to 7.5 kW for 230 V systems or 13 kW for 400 V systems. Protection by obstacles, placing out of reach, non-conducting location and earth free local equipotential bonding cannot be applied. A grid with a spacing of not more than 30 mm must be placed above a floor-heating element or below a ceiling-heating element and connected to the protective conductor of the system.

To avoid the temperature of floor or ceiling heating systems exceeding the required maximum of 80°C, the design and installation must be appropriate; it is recommended that floors that may be used by people not wearing substantial footwear should not exceed 35°C. These requirements can usually be met by careful design in conjunction with the manufacturer of the heating system, although temperature sensing protective and control devices may be used. The heating units must be connected to the electrical installation using cold leads or terminal fittings, which must be connected to heating units by welding, brazing or compression techniques. Where expansion joints are used in a construction, heating units must not cross them.

Flexible heating elements must comply with BS EN 60335-2-96 and heating cables with the BS 6351 series. Where installed in ceilings, heating units must have protection to at least IPX1, or to IPX7 in floors of concrete or similar materials.

8.23.3 Identification [753]

These heating systems are subject to very detailed requirements as far as paperwork is concerned. The installer must provide a detailed plan of each system, which must be fixed on, or adjacent to, the switchgear concerned. In addition, the installer must provide a very full description of the heating system, including its construction, the depth of buried heaters, the exact locations of heaters and of heating-free areas, the control equipment with circuit diagrams and details of the maximum operating temperature.

The installer must provide as many copies as required to ensure that all who use the system have access to the information. One copy should be fixed in or near each distribution board. Instructions for use must include

1 the manufacturer and the type of heating units
2 the number of heating units installed
3 the length and/or area of the heating units
4 the rated power
5 the surface power density
6 a drawing to show the layout of the heating units
7 the position and depth of heating units
8 the positions of junction boxes
9 details of conductors, shields and the like
10 the heated area
11 the rated voltage
12 the rated cold resistance of the heating units
13 the rated current of overcurrent protective device
14 the rated residual operating current of the RCD(s)
15 the insulation resistance of the heating installation and the test voltage used
16 the leakage capacitance

This information should be displayed close to the distribution board of the heating system and form part of the Electrical Installation Operational Manual (see {7.8.1}).

8.23.4 Bathrooms and swimming pools
A 30 mA RCD that has no time delay must also protect these floor or ceiling heating systems. Such installations are likely to suffer from the ingress of water and from corrosion, so terminals must be easily accessible to facilitate testing. Particularly in swimming pools, regular testing and inspection should be carried out and adequate records must be kept. (See also {8.2.2 and 8.3.2}).

8.24 EXTRA-LOW VOLTAGE LIGHTING INSTALLATIONS
8.24.1 Introduction
Extra-low voltage lighting has become so widespread over many years that it is, perhaps, surprising that BS 7671 does not mention it specifically; this section is based on Chapter 15 of Guidance Note 7. All wiring, including that at 12 V and 24 V is, of course covered, but no specific mention has been made of the extremely common use of very small tungsten halogen lamps now in very wide use, particularly in commercial premises such as shops, hotels, etc. BS 7671 still does not mention this important part of many installations, but the presence of draft CENELEC standard HD 384.7.715 and IEC 364.7.715 has resulted in Chapter 15 on the subject in Guidance Note 7. These drafts cover lighting installations fed at up to 50 V a.c. or 120 V d.c.

8.24.2 Basic requirements [414, 43]
Only fixed SELV systems may be used. All these circuits operate at low voltage derived from the mains supply using transformers, which must be to BS 3535 and BS EN 61558-2-6, or using electronic voltage converters to BS EN 61046 and with

BS EN 60598-2-23. Transformer secondary windings may only be connected in parallel where the primary windings are similarly connected, and where the transformers have identical characteristics. Overcurrent protection must be the same as that for other circuits [Chapter 43], although account must be taken of the inrush current to the transformer when switched on. The maximum disconnection times in final circuits supplying up to 32 A at up to 50 V AC are 0.8 s for TN systems and 0.3 s for TT systems. The overcurrent protective device must be of the non self-resetting type, and in its selection account must be taken of the transformer magnetising current.

Extra-low voltage lighting generates heat, as do all filament lamps; their transformers may also get hot. Care must be taken to ensure that lamps and transformers are not installed where they could ignite flammable materials. Where the nature of processed or stored materials renders them particularly prone to fire, the power demand of the SELV circuits must be continuously monitored and arranged for the supply circuit to be disconnected within 0.3 s if the power increase exceeds 60 W.

8.24.3 Wiring systems [414]

The low voltage wiring systems follow the basic rules of BS 7671. Extra-low voltage wiring may be conventional, but may also be in other forms, such as track systems to BS EN 60570. Metallic parts of a building, such as pipe systems, may not be used as live conductors. Conductors must never be used as supports for notices, a very common occurrence in shops. Where parts of the system are accessible, people must be protected from burns by ensuring that surface temperatures do not exceed those given in {Table 3.1}.

The minimum size of conductor for these systems is 1.5 mm², although 1.0 mm² flexible cables may be used provided that they are no longer than 3 m. It must be remembered that a lamp of a given power takes more than nineteen times as much current at 12 V as it does at 230 V. Bare conductors may be used, but only for voltages of up to 25 V a.c. or 60 V d.c., but must have a c.s.a. of at least 4 mm² (for mechanical strength) and must not be in contact with combustible materials; where they are suspended above such materials, there should be a clearance of at least 300 mm.

8.25 MOBILE OR TRANSPORTABLE UNITS

8.25.1 Introduction [717.1]

Mobile or transportable units are defined as vehicles and/or transportable structures in which part or all of a low voltage electrical installation is contained. 'Units' may be mobile or transportable, self-propelled, towed or moved by other means. Examples of transportable units are

− broadcasting vehicles
− medical services vehicles
− fire fighting appliances
− mobile workshops
− transportable catering units

Excluded from these requirements are
1 transportable generating sets
2 marinas and pleasure craft (see {8.18})
3 mobile machinery
4 caravans and leisure accommodation vehicles (see {8.10})
5 traction equipment for electrical vehicles

8.25.2 Problems and solutions

There are many possible problems with transportable units, which include

a) the loss of earth connection as a result of using temporary cables, long supply runs and wear and tear

b) the possibility of connection to differing supply systems having unusual characteristics

c) the impracticability of establishing an equipotential zone externally

d) open circuit faults on the PEN conductor of PME supplies (see {5.6})

e) electric shock due to high protective conductor current where the unit contains large amounts of electronics equipment

f) faults caused by vibration during movement

These risks can be reduced by

1 making sure of the suitability of the supply before connection

2 installing extra earth electrodes where necessary

3 introducing a log book to record regular inspection of connecting cables and their connectors

4 always using stranded or flexible cables of cross section 1.5 mm² or greater for internal wiring

5 supply cables must be stranded or flexible and of cross section 2.5 mm² or greater

6 users outside the unit should be protected by the installation of 30 mA RCDs

7 using RCDs to provide protection against fault contact

8 where possible, using an isolating transformer or an on-board generator to provide electrical separation

9 where possible, using earth-free equipotential bonding (see {5.4.3})

10 where possible, using Cass II enclosures (see {5.8.1})

11 clear labelling, especially to indicate the types of supply which may be used

12 instituting a robust system for regular inspection and testing

8.25.3 Detailed requirements [717]

The information provided by Section 717 of BS 7671 is extremely extensive and detailed. So much so, that it is felt to be beyond the scope of this book to describe it fully. Any electrician who is required to deal with equipment of this kind is strongly advised to obtain and study Section 717 of BS 7671, Guidance Note 7, Chapter 17, and/or draft IEC standard 364-7-717/Ed1.

CHAPTER NINE

Alarm systems

9.1 INTRODUCTION

There is a very clear need for alarm systems, which are of two distinct types. They help us to protect our property and our lives against the twin dangers of fire and of unauthorised intruders. These two types of alarm system have much in common. In both cases some person or event must trigger the alarm, and in both cases the result is a warning, although in the case of an intruder alarm system, this may be silent and not obvious to the intruder. Both will usually require hard wiring, although, particularly in the case of intruder alarm systems, wireless types relying on radio signals are becoming common. In both cases there is likely to be a control panel, and in many situations the alarm sounders may be similar.

Fire alarms are necessary to alert those using a location of the outbreak of fire so that they may evacuate from the area safely. Because people are in physical danger where there is a fire, considerable legislation applies to fire alarm systems. For example, they are a legal requirement in most commercial and industrial situations, their installation and maintenance being subject to a large number of safety regulations. The law demands that careful records are kept covering installation, maintenance and regular testing.

Intruder alarms have become commonplace as a means of protecting our valuables from those who seek to steal them. Whilst equally important, some may say more important, than fire alarms, they are not surrounded by the same legal requirements. There is no law to say how they should be installed, tested or recorded. However, authorities, such as insurance companies and the police are very likely to have detailed requirements before they will underwrite an insurance policy or agree to attend in the event of an alarm being triggered. Whilst not being enforceable in law, these requirements may well be just as onerous as those for fire alarms, the difference being that the penalties will be financial rather than judicial.

There is thus something to be gained by treating both types of alarm together, as if they were common systems. However, the differences, particularly in the legal requirements, make this approach difficult. Every attempt has been made to avoid repetition, but in a few cases this has not proved to be practicable. The standard for alarms, BS 4737, has been superseded by document PD 6662: 2004. This is not a standard as such, but includes European Standard EN 5031-1 2004.

9.2 FIRE ALARM SYSTEMS

All wiring must, of course, comply with BS 7671: 2008, the 17th Edition of the Requirements for Electrical Installations. Section [422], 'Precautions where particular risks of fire exist', and Chapter [55] 'Safety services', apply to these systems. As mentioned earlier, fire and intruder alarms have much in common. However, because of the more onerous legal requirements for fire alarm systems, they are treated here as separate systems.

9.2.1 Domestic smoke alarms

Unless an automatic fire detection and alarm system is fitted, domestic premises must be provided with mains operated smoke alarms to BS 5446. Further guidance can be found in BS 5839. There are two types of domestic smoke alarm.

1 Ionization chamber type, which is better at detecting fast burning fires and which is recommended for installation in living or dining rooms.

2 Optical types, responding better to smouldering fires and which should be installed in situations such as hallways and landings.

Mains-operated smoke alarms may be fitted with a secondary power supply such as a replaceable or rechargeable battery or a capacitor. There should be at least one smoke alarm for each storey of a dwelling, and they must be linked so that smoke detection by one operates them all. Smoke alarms should be ceiling mounted, at least 300 mm from walls and light fittings, and there should be an alarm within 7.5 m of the door to every habitable room. They should not be fixed next to or immediately above heaters or air conditioning units. Alarms designed for wall mounting may be used provided that they are above the level of doorways and fixed in accordance with the manufacturer's instructions.

Smoke alarms should not be fitted in extreme temperatures, such as in boiler rooms or exterior porches, or in bathrooms, showers, cooking areas or garages. They must be positioned so as to be easily accessible for testing and cleaning, so it is unwise to mount one over a stair shaft.

The power supply for mains operated smoke alarms should be from a separate circuit, and if the alarms do not include a stand-by power supply no other equipment should be fed by it other than a dedicated mains failure alarm. Where standby power is available, the alarms may be connected to another circuit such as one used for lighting. The circuit should preferably not be protected by an RCD, but if one is essential, either

1 the smoke alarm circuit should have its own dedicated RCD, or

2 the RCD should operate independently of any other RCD.

There is no requirement for the cables feeding smoke alarms to be different from those used in the rest of the installation or to be segregated.

9.2.2 Detection devices

The object of these devices is to signify the presence of a fire to the alarm sounders, sometimes via a control panel. A number of alternative devices is available.

1 Break-glass call points

These devices, sometimes called manual detectors, allow a person to raise the alarm in the event of a fire. In some cases a hammer attached by a chain is made available, whilst in others a thin glass cover is scored so that pressure from a thumb will break it. A protective plastic coating prevents glass fragments from injuring the operator or from jamming the operating switch. A key for test purposes can operate this type.

2 Automatic detectors

These devices must be able to differentiate between the presence of a fire and the normal environment of the area concerned. It is of vital importance that automatic detectors are correctly chosen and properly sited. For example, smoke detectors should never be used where they could possibly be subjected to smoke or fumes from a kitchen or from a boiler since false alarms will result. It must be borne in mind that the greatest concentration of heat and of smoke will occur at high levels, where detectors should be mounted. There are many forms of automatic detector, including:

3 Ionisation smoke detectors

These devices respond when smoke or fumes enter the detector chamber resulting in ionisation of the gas inside. As well as visible or invisible smoke, they respond to various chemicals, including petrochemicals. They are most effective when the fire to be detected is a flaming, rather than a smouldering, conflagration.

4 Photoelectric (optical) detectors

A light emitting diode (LED) radiates to a photodiode, producing an electrical signal. The presence of smoke scatters and reduces the light received by the photodiode, changing the electrical signal and triggering an alarm. This detector is suitable where smouldering fires are more likely.

5 Beam detectors

These types are similar to optical types, but have a light transmitter separate from the receiver. The two components may thus be widely separated, and hence are suitable for fire detection in wide, open spaces.

6 Heat detectors

Temperature is measured by a thermistor network, which provides an output signal depending on external temperature. Some types are programmed to give an output signal only when the rate of rise of temperature exceeds a preset level, unless temperature becomes excessive.

7 Flame detectors

This type is sensitive to low frequency flickering infra-red radiation emitted by flames during combustion.

9.2.3 Control panels

The control panel will vary depending on the complexity and the size of the alarm system. Most industrial types are computerised, and will give direct indication of the specific area in which the fire is located. Many will have a radio or telephone connection to give early warning to the local fire station. They must comply with BS 5839: 1988 'Fire Detection and Alarm Systems for Buildings' and must be provided with standby batteries so that operation is ensured in the event of mains electricity failure.

Some panels will have an automatic fault indicator, as well as a system for showing the integrity of the mains supply, the battery charging system and the battery condition. The panel must be installed in an area of low fire risk, usually on the ground floor. Some panels are equipped with door release facilities; fire doors, kept open for ease of access during normal times, are released and spring shut in the event of an alarm.

9.2.4 Fire alarm sounders

These devices are required to give warning of fire after its detection. They are some-times electro-mechanical bells, but may be electronic sounders. In all cases the sound must be distinct from that of other sounders (such as intruder alarms) and must be loud enough to ensure that they can be heard. A minimum sound level of 65 dB, or 5 dB above background noise is usually required. Sounders must be wired on at least two separate circuits to ensure that failure of one will not prevent the sounding of an alarm.

9.2.5 Types of fire alarm systems

BS 5839: 1988 'Fire Detection and Alarm Systems for Buildings' covers the design, installation and maintenance of fire detection and alarm systems.

The BS divides fire detection and alarm systems into eight categories. These cate-gories are

Type P   Automatic detection and alarm systems for the protection of property only
Type P1   Detection and alarm systems installed throughout the protected buildings
Type P2   Detection and alarm systems installed only in defined parts of the buildings
Type L   Automatic detection and alarm for the protection of life
Type L1   Detection and alarm systems installed throughout the protected buildings
Type L2   Detection and alarm systems installed only in defined parts of the buildings
             NOTE that a Type L2 system should also include the coverage required for a Type
             L3 system (see below).
Type L3   Detection and alarm systems installed only for the protection of escape
Type M   Manual alarm system. There is no sub-division.

BS 5839 inserts the letter 'D' to designate systems installed in dwellings. Thus an automatic system for the protection of life and installed in a block of flats will be shown as Type LD2.

It is important that the escape procedures to be followed by operation of the alarm are known before an alarm system is designed. Such design must take account of how the evacuation will be conducted, including the possible need for a secure public address system to give advice. It must be noted that a Type L3 system is installed only for warning on an escape route, the system itself is not expected to protect people who are fighting the fire.

9.2.6 Power supplies and cables

Each fire detection and alarm system must be connected to the mains supply by its own isolator, which must be coloured RED and must be labelled:

> **FIRE ALARM: DO NOT SWITCH OFF**

To prevent unwanted operation, the isolator may be contained in an enclosure with a breakable glass cover. If the isolator is fed from the live side of a supply so that the main switch does not switch it off, there should be a second label, which reads:

> **Warning. This supply remains live**
> **when the main switch is turned off**

In addition the main switch must carry the label:

> **Warning. This fire alarm supply remains live**
> **when this switch is turned off**

Alternatively, if the main switch directly controls the fire alarm system, it should carry the following label:

> **Warning. This switch also controls the supply to the fire alarm system**

The operation of a fire alarm system is entirely dependent on the integrity of the cables connecting its various components, such as alarm sounders, supply system, bells, sirens and so on. An important part of the design is thus to ensure, as far as possible, that the cables will remain in working condition should a fire occur. It follows that the alarms should continue to sound during a fire, even if the supplies to detectors or other points of initiation have been destroyed. This means that the cabling used for the alarms should be able to continue to operate for at least half an hour after the event of fire.

The cables used should therefore be either:

1 mineral insulated copper sheathed cables to BS 6207, or
2 cables complying with BS 6387 and meeting category requirements AWX or SWX.

Other types of cable may be used provided that they are protected from the fire by being buried under at least 12 mm of plaster, or are separated from the fire risk by a wall, partition or floor having a resistance to fire for at least half an hour.

As well as having good fire performance as indicated above, it is sensible to segregate alarm cables from others so that they will not be damaged in the event of faults on the other circuits. Generally speaking, fire alarm circuits should be separated from all other electrical wiring by one of the following methods.

1 Using a dedicated conduit, trunking, ducting or channel system for fire alarm cables
2 Separating alarm cables from others using non-combustible partitions which are rigid and continuous
3 Using mineral insulated copper cables having an insulating sheath
4 Using cables complying with BS 7269 which have metallic or insulating sheaths
5 Ensuring that there is a space of at least 300 mm from other cables.

The wiring of an alarm system is required to have the same regular inspection and testing as other fixed installations. Intervals not exceeding five years are specified for periodic test and inspection.

9.2.7 Record keeping

Drawings and operating instructions for the alarm system must be maintained and should ideally be kept close to the control and indicator equipment.

One person should be ultimately responsible for the system, and should maintain a log book which includes:

1 His own name, location and telephone extension number
2 Full details of the servicing arrangements
3 A complete record of all alarms including the location from which the alarm was raised. The list must include practice, false, test and genuine alarms
4 The dates, times and nature of every defect or fault that has occurred
5 Full data, including dates and times, of when these defects or faults were rectified

6 The date and time of each service of the system
7 The dates and types of all system tests
8 The dates and types of all system services
9 The dates and times when the system was disconnected or non-operational
10 Full details of alterations to the system, including dates
11 Full details of the people and/or organisations carrying out work of any kind on the system.

The layout of a typical logbook follows.

A responsible executive should maintain this book, making sure that every event is properly recorded.

Events will include:
1 Real alarms
2 False alarms
3 Pre-alarm warnings
4 Tests
5 Faults
6 Temporary disconnection
7 Visits by installing or servicing engineers, with data concerning the work performed on each visit

Name of the responsible person ....
Address ....
Telephone number and extension of the responsible person ....
System installed by ....
Maintained by ....
Date of cessation of maintenance contract ....
Telephone number of servicing contractor ....

Date	Time	Event	Action	Initials

The responsible person must have the authority to ensure that:
1 the detection and alarm system is correctly serviced and maintained
2 all staff are instructed and trained in the actions they should take in the event of an alarm being initiated
3 the local Fire Authority approves of the escape plan
4 all escape routes and access to fire alarm and extinguishing equipment are kept clear at all times

9.2.8 False alarms

False alarms will not only waste valuable time, but may also result in a loss of confidence in the system, which could have serious results. Notices should be widely displayed to draw attention to the system. It is particularly important to ensure that automatic alarm dialling systems to the Fire Service (999 or 112) are disconnected before a routine test. Staff, both those working on the site and those visiting it, should be made aware of the alarm system and of the precautions to be taken if false alarms are to be avoided. Where engineering or building work is undertaken on the site, those concerned must be given advice so that they take precautions to avoid damage, disconnection, or false operation of the alarm system.

9.2.9 Servicing

Servicing is likely to be best arranged on a regular basis. All defects that come to light must be entered in the logbook. Details of the checks and work required at each regular service follow.

Daily service

1 The alarm panel indicates normal operation – if there is a fault, check that this has been recorded
2 Unless there is continuous monitoring, test the automatic link to the fire brigade
3 Check that the faults recorded yesterday have been rectified

Weekly service

1 At least one alarm initiator (break glass point, automatic sensor, etc) on each zone must be tested to verify the ability of the system to receive the signal and to sound the alarm
2 Visually inspect the batteries and their connections where possible. Report and rectify defects
3 Where prime movers (petrol or diesel engines) are involved, check fuel, lubrication and coolant
4 If the system has no more than thirteen zones, at least one zone must be tested each week. Where there are more than thirteen zones, more than one zone must be tested every week to ensure that each is tested at least every thirteen weeks
5 Enter any defects detected in the log book

Monthly service

1 Where automatically started standby generators are used, they must be operated by a simulated failure of the main power supply system and allowed to run for one hour. Checks should be carried out to ensure that the operation of the generator(s) has not affected them
2 After testing, fuel, lubrication and coolant levels must be checked and adjusted.

Three-monthly service

1 Check that all faults shown in the logbook have been rectified
2 When a detector in each zone is operated, check that the alarms and indicators work correctly
3 All ancillary functions of the control panel must be tested
4 Simulation of appropriate faults is used to check all fault indicators
5 All batteries are examined and checked, with necessary measurement of voltage and, where appropriate, of specific gravity of the electrolyte
6 A complete visual inspection of the building should be carried out to ensure that no changes in structure or in use have taken place which may require alteration to the alarm system layout

Annual service

1 All cables and fittings must be visually inspected to ensure that they are secure
2 All equipment must be checked and examined for signs of ageing, of wear or other indications of possible malfunctions
3 Every detector must be checked for correct operation

9.2.10 Certification

A new detection and alarm system should be certified on completion of inspection and testing with a certificate such as the following:

Fire Alarm System Certificate of Installation and Commissioning
For the user
Area protected ....
Occupier ....
Address ....
I confirm that I have been made aware of BS 5839 and in particular to clauses 14 (false alarms), 28 and 29 (user responsibilities)
Signed .... Status .... Date ....
For and on behalf of the user ....

For the installer
The installation has been inspected and tested to BS 5839 standards for insulation resistance of all conductors, and earthing. The operation of the entire system has been tested and it is certified that it complies with all the requirements of BS 5839 other than the following deviations.

....
Signed (Commissioning engineer) ....
Name in block capitals ....
For and on behalf of (installer) ....
The system log book is situated ....
The system documentation is situated ....

For periodic testing of a fire alarm system, the following certificate is appropriate.

Certificate for Testing a Fire Alarm System
This certificate relates to the testing of the fire alarm system at
Protected area ....
Name of occupier ....
Address ....
The system has been inspected and tested to the requirements of BS 5839.
Work carried out and covered by this Certificate is as follows.
(Items not covered are deleted).
Alterations and extensions to the existing system
Quarterly inspection and test
Annual inspection and test
Servicing following a fault
Servicing following a pre-alarm warning
Servicing following a false alarm
Servicing following a fire
Other non-routine service (specify)
....
Signed .... Status .... Date ....
Name in block letters .... Acting for ....

9.3 INTRUDER ALARM SYSTEMS

As mentioned earlier, intruder alarms and fire alarms have much in common. However, because of the more onerous legal requirements for fire alarm systems, they are treated in this book as separate systems. Security and protection for homes and businesses is becoming an increasing problem, as a result of which intruder (often called burglar) alarm systems have proliferated. Whilst intruder alarms are sometimes integrated with fire alarm systems, this is by no means common.

9.3.1 Detection devices

The purpose of the detection device is to detect the intrusion and to provide a signal resulting in a warning. Devices are in two categories.

1 Perimeter devices are intended to identify unauthorised entry, and consist of door and window operated types.
2 Motion detectors sense movement in areas where there should be no people.

There are many types of detectors, including:

1 Magnetic reed switches

Perhaps the most common protection device, the magnetic reed switch consists of a pair of normally closed contacts held in that position by an adjacent magnet. When the magnet moves away, following the opening of a door or a window, the contacts open, initiating the alarm. These switches can be recessed in doors, windows, etc. so as to not be apparent to the casual observer.

2 Impact and vibration detection devices

The device relies on a piezo-electric crystal that produces electrical impulses when subjected to impact or vibration. They are fitted to doors, windows, roofs, safes and so on. They protect against, for example, an intruder who breaks through the panel of a door to gain access, without actually opening it.

3 Break-glass (acoustic) detectors

This type may be fixed to the surface of glass and responds to vibration when the glass is broken. Alternatively, the device may be triggered by the sound of breaking glass. This type may be fitted to the window frame or to the ceiling at a distance from the window itself.

4 Passive infra-red (PIR) detectors

This is the most common of the motion detectors. The device receives and monitors all of the infra-red (heat) energy radiated from all objects within its field of view. Any change in energy large enough or rapid enough will trigger the device. PIRs are made with a wide variety of optical coverage patterns, so that they can be chosen for the particular situation concerned. Thus, for example, an area could be covered except for a short distance above the floor, to prevent false alarms due to the presence of pet animals. To cover a space and to give reliable operation, and intruder-detection PIR is best mounted in a corner position at ceiling height, but out of direct sunlight.

5 Ultrasonic detectors

This type operates by transmitting a high frequency sound signal, above the upper frequency response of the human ear. The signal received will have been reflected from all surfaces within the space covered, and provided that these surfaces are stationary, will be at the same frequency as that transmitted. If, however, anything within the space moves, the Doppler effect results in the received signals being at a different frequency from those transmitted and the alarm is triggered. The ultrasonic detector is larger than the PIR and the range covered is usually limited to about 8 m.

6 Combined ultrasonic/PIR (dual technology) detectors

Combining the two types of detector into one unit will sometimes result in a very low level of false alarms.

7 Microwave detectors

In some ways these units are similar to ultrasonic types, but instead of emitting high frequency sound they employ microwaves. This increases the operating range, often up to 30 m, the effective range depending on the size of the target (intruder). They are often used to trigger closed circuit television recording systems in outdoor situations.

8 Combined acoustic/PIR detector

The combination of acoustic and PIR detection into a single unit can result in very low false alarm figures. A logic system ensures that the alarm will only be triggered if the acoustic (break glass) activation is followed by movement detection by the PIR.

9 Active infrared (AIR) detectors

An infra-red beam is transmitted across the area to be protected, and is focussed onto a receiver unit. If an intruder breaks the beam, the alarm is triggered. These systems are unaffected by direct sunlight, and enable large areas to be protected.

10 Capacitance detectors

The presence of a foreign body close to a metal surface will alter its capacitance, and this effect is used to trigger an alarm. It is used to protect safes and metal filing cabinets with valuable contents. Metal foil strips can be fixed to non-metallic objects to enable capacitance detector protection to be provided.

9.3.2 Control panels

There is a very wide range of control panels available to suit most specialist needs. Many have keypad controls to enable the user to select the alarm zones of a building to be activated or deactivated as required. Time delays are used to allow the operator to leave the alarmed area before the system begins to operate, as well as to allow entry for switch-off. In most cases the alarm will operate if the wrong identification code is entered.

Most panels are provided with fail-to-safe systems, which will ensure that if the wiring system for the alarms is subject to tampering, the alarm is triggered. Many will indicate the zone of the area protected in which the alarm was initiated; in some cases, a log record will be kept automatically to show the zone concerned as well as the time at which the alarm was triggered.

9.3.3 Alarm sounders

Electro-mechanical bells or electronic sirens are often used to draw attention to activation of the alarm system. In some cases, warning strobes or beacons provide a visual alarm. The sounders or visual displays are themselves often provided with sensors so that tampering with them will raise the alarm. The sounders are often operated from a battery system so that their alarm is not lost in the event of mains failure. In some cases, there may be no audible or visual alarm to indicate to the intruder that he has been detected. Instead, an automatic telephone or radio call is made to the police to warn them of the intrusion.

9.3.4 Wiring intruder alarm systems

Unlike fire alarm systems, there are no legal requirements for the wiring of intruder alarm systems. Mains wiring must comply with BS 7671:2008 (the 17th Edition of the IEE Wiring Regulations) but there are none of the onerous conditions applying to fire alarm systems.

However, in some cases the contents insurance policy for premises may require that certain special requirements be complied with. Occasionally, the security organisation concerned or the police will also stipulate certain minimum standards before they are prepared to agree to respond to alarms. Since insurance company requirements will differ from company to company, as will police requirements from different forces, no clear guidance can be offered here. It is, however, important that the installer should ascertain the special requirements for each installation, and should make sure that his work complies with them.

9.3.5 Servicing and certification

As with wiring systems, there are no overall rules governing the need for regular servicing or for the maintenance of servicing and testing records. Again, the needs for individual installations are likely to be set by insurance companies, security firms or the police concerned, and it is important for the users to comply with these requirements. Failure to do so will not result in legal action (as can happen with fire alarm systems) but could lead to insurance claims being refused or to failure by the police or the security organisation to respond to alarms.

CHAPTER TEN

Emergency lighting

10.1 INTRODUCTION

Perhaps the prime purpose of emergency lighting is to allow the occupants of a building to escape along a safe route in the event of the loss of the primary lighting system. This may well be as the result of a fire or other event which could quickly become a disaster if escape is hindered in any way. These routes must be clearly marked and adequately lit to offset the panic that often follows the loss of full artificial lighting in the event of a fire.

Another very important function of emergency lighting is to allow staff in switch and control rooms to continue their work in the event of the failure of the main lighting system This may be to restore supplies or to allow dangerous plant to be shut down. Such systems are sometimes described as standby lighting.

All wiring must, of course, comply with BS 7671: 2008, the 17th Edition of the Requirements for Electrical Installations. Section [422], 'Precautions where particular risks of fire exist', and Chapter [56] 'Safety services', apply to these systems.

10.2 ESCAPE ROUTES AND EXITS

The Fire Precautions (Workplace) Regulations: 1997 require the provision of escape routes and exits. The number of escape routes and exits to be provided will depend on

1 the number of occupants in the room or the building
2 the size of the room and of the building
3 the height and number of storeys (levels) of the building
4 the distance to the nearest exit.

These routes are defined in Approved Document B of the Home Office's guidance to fire precautions in places of work that require a fire certificate. The routes, which must be kept unobstructed at all times, are usually marked with 'tram lines', and escape doors must never be locked whilst the building is occupied. Alternative routes may be necessary in some cases. Escape routes must be provided with emergency lighting and with signs to indicate the safe direction of travel.

10.3 SAFETY SIGNS

The European Council directive 92/58/EEC is implemented by the Health and Safety (Safety Signs and Signals) Regulations 1996 in specifying the minimum requirements for emergency escape and fire safety signs in the workplace. BS 5499 and EN 1838 also apply. These signs should be clear in all circumstances, and with this in mind they are required to be mounted at between 2 m and 2.5 m above floor level. A formula is given from which the minimum size of the sign can be calculated; since electricians are unlikely to provide signs, details are omitted.

Escape route, exit and emergency exit signs must be illuminated so that they are always legible. In the event of a failure of the main lighting system, they are required to remain illuminated by the emergency lighting system. This illumination may be from

1 an internally-illuminated sign, which should comply with BS 5499 Part 1, or
2 an external emergency lighting luminaire.

Sign Colours

The colours of signs must comply with the requirements of 92/58/EEC, which are shown in Table 10.1.

Table 10.1 Colours required for safety signs

Colour	Purpose	Information or instruction
Red	Prohibition sign	Dangerous behaviour
Red	Fire fighting equipment	Identification of location
Red	Danger alarm	Stop, shutdown & emergency cut-out devices
Green	Emergency escape & first aid signs	Doors, exit routes & equipment facilities

Fig 10.1 Emergency exit or escape route sign

- The arrow may point the applicable direction
- The rectangle, the figure and arrow are white on an olive green background

10.4 ESCAPE LUMINAIRE SITING

An emergency luminaire must be placed in every position where it will help persons to escape or where it will facilitate operation of the alarm system. The mounting height of emergency luminaires is important. If too low, especially in a corridor, they may be obscured by the movement of persons and be subject to damage. If too high they may be obscured by a layering of smoke in the event of fire.

The following positions for the siting of escape luminaires are a requirement of BS 5266.

1 At all exit doors used in the event of an emergency
2 At changes in direction of the escape route
3 At the intersection of corridors
4 At or near changes of floor level
5 On staircases, so that each flight of stairs receives direct light
6 Close to every fire alarm call point
7 Near fire fighting equipment
8 At all safety signs, including exit signs
9 Outside final exits

The same British Standard also recommends that emergency luminaires should be provided in the following locations.

10 Moving stairs and walkways
11 Lift cars
12 Outside all exits
13 Toilets and similar situations which exceed 8 m^2 in area

10.5 INSTALLATION OF EMERGENCY LIGHTING

Emergency lighting should not be considered as a separate issue to the main lighting, but as a special part of it. In other words, emergency lighting should be an integrated part of the total lighting, not an add-on extra. Filament lamps may be used, but suffer from low efficiency and short life. Fluorescent and compact fluorescent lamps are the better choice, and light emitting diodes (LEDs) are becoming very common due to their robust qualities and very long operating life. High pressure discharge lamps are not usually employed because of their prolonged starting times.

Two basic types of supply are in use for emergency lighting.

1 Self-contained types, which take their supply from the mains when it is available but switch to an internal battery on mains failure. The battery is charged by the mains supply when it is available, and on mains failure will either directly feed a filament lamp or, via an inverter, feed a fluorescent lamp.

2 Central power type, where a central supply from batteries or a generator feeds the luminaires. There are three available types

   a) AC/DC battery systems supply direct current from the batteries, usually at 24 V, 50 V or 110 V, to the luminaires.

   b) DC/AC battery systems provide a 230 V output from the batteries using an inverter, or the battery is used to start a generator that provides a similar supply.

   c) Uninterruptible power supplies (UPS) continue to provide their output without a break following mains failure enabling their use with discharge lamps that would otherwise have unacceptably long re-strike times. Both BS EN 50091 and BS EN 50171 must be followed for these devices. The charger used must be capable of recharging the battery to 80% of its capacity within 12 hours, and must be designed for a life of 10 years. Lower life batteries have a sudden failure mode that will not be picked up by emergency lighting testing procedures.

Wiring to self-contained luminaires is not considered to be part of the emergency lighting circuit, since its integrity does not affect the operation of the luminaire in the event of mains failure. All other wiring to emergency luminaires must be protected from physical damage and from fire. All such cables must be routed through areas of low fire risk. Wiring should either be in mineral insulated cables to BS 6207, or cables that comply with category B of BS 6387. Such cables are either those with thermosetting insulation to BS 7629 or armoured types to BS 7846. Other cables, such as PVC insulated types in steel or rigid PVC conduits are permissible, provided that they are protected from fire. Such protection must be by burying the cable in the structure of the building, or running them in situations of minimal fire risk and where a fire-resistant wall, floor or partition protects them.

Supplies to emergency lighting

It is important to appreciate that where the supply is switched off at night, this may, unless special arrangements are made, prevent the charging of a central battery or of those in self-contained luminaires. It will be appreciated that when the central supply is switched off, the emergency lighting will see this as a mains failure and come into operation, discharging its batteries in the process.

Isolation of emergency lighting

Special precautions should be taken to avoid unexpected switching of the supply to an alarm system. The required label should read

> **Emergency/Escape/Standby lighting**
> **Do NOT switch off**

Supplies for safety services are the subject of Chapter 56 of BS 7671: 1992 'Requirements for Electrical Installations'.

10.6 MAINTENANCE OF EMERGENCY LIGHTING

When an emergency lighting system is completed, the installer will provide a completion certificate. Maintenance of the system should be carried out on a daily, monthly, six-monthly and three-yearly basis as in the following description. All these maintenance activities must be recorded in a logbook. The logbook must record all of the following information. Increasing use is being made of computer-based self-testing systems.

1 The date of the completion certificate, and of certificates for extensions and alterations
2 Dates and details of every service, inspection and test of the system
3 Dates and details of all reported defects, together with details of remedial action taken
4 Dates and details of all alterations to the emergency lighting system

Details of the actions required at each of the periodic maintenance events are as follows.

Daily service

1 Check that faults already entered in the logbook have been removed
2 Check that lamps in all maintained luminaires are lit
3 Record any faults found in the logbook
4 Check that the main control panel of central batteries or engine-driven plant shows a healthy indication

Monthly service

Some of the tests involve a simulated failure of the mains supply. The period of simulated failure must not exceed 25% of the rated duration of the standby batteries to ensure that they are not depleted. Where the required checks cannot be made in this time, the process must be repeated after the batteries have been fully recharged.

1 Simulate a failure of the mains supply for long enough to check that all self-contained and illuminated exit signs are operating.
2 A similar simulation must be carried out to check the correct operation of a central battery system.

3 This simulation must be applied to test engine-driven generating plants, which must energise the emergency lighting for at least one hour.

4 Fuel levels, as well as oil and coolant supplies on engine-driven generating plant, must be checked and adjusted.

5 For engine-driven back-up plant with automatic starting using batteries, simulated mains failure must be carried out with the engine prevented from starting. After this check, normal starting must be allowed on simulated mains failure, with the generating plant energising the emergency lighting for at least one hour.

Six-monthly service

The full monthly tests should be carried out, plus operation of three-hour rated luminaires, signs and back-up batteries for one hour. Where these devices have a rated duration of only one hour, the test should be for fifteen minutes. No mention is made of two-hour systems, but common sense would suggest that they should be tested for thirty minutes. The Control panels must be checked for healthy indications.

Three-yearly service

The full monthly inspection should be carried out. In addition, every self-contained luminaire, internally-illuminated sign, central battery system and generator back-up battery must be tested for its full duration. The complete installation must also be examined to confirm that it still complies with its stated form.

Additionally, it is required that self-contained luminaires with sealed batteries that are more than three years old should be subjected to the three-year test every year to ensure that they have adequate storage capacity.

10.7 DURATION OF EMERGENCY LUMINAIRE OUTPUT

There are two types of emergency lighting luminaire.

1 Maintained emergency lighting – given the designation M. This is a system where all emergency lighting is in operation at all times.

2 Non-maintained emergency lighting – given the designation NM.This system is one in which emergency lighting is only operational when normal lighting fails.

The number of hours for which both types can maintain their light output to an acceptable level after the supply failure is added to the designation. For example, designation NM/2 indicates a non-maintained luminaire with duration of two hours. Evacuation is unlikely to take longer than one hour, but Local Authority licences sometimes require extended periods.

The minimum requirements for duration are shown in Table 10.2 overleaf.

Table 10.2 Emergency lighting duration requirements

Type of Premises			Minimum Category
Residential	With sleeping accommodation, such as hotels, nursing homes, guest houses, hospitals, educational establishments with boarders	10 bedrooms or more	NM/3 or M/3
		Small – less than 10 bedrooms and with not more than one floor above or below ground level	NM/2 or M/2
Non-residential	Care or treatment, such as clinics, special schools, and so on.		NM/1
Non-residential	Recreational such as public houses, restaurants, concert halls, sports halls, exhibition halls, theatres, and so on.	Buildings where lighting can be dimmed or for the consumption of alcohol	NM/2
		Other	NM/2 or M/2
		Where not more than 250 people are present	NM/1 or M/1
Non-residential	Recreational such as cinemas, ten pin bowling venues, bingo halls, dance halls, ballrooms, and so on.		As non-residential immediately above, but subject to BS CP 1007 for cinemas
Non-residential	Laboratories, colleges and schools		NM/1 or M/1
Non-residential	Industrial, such as manufacturing, storage, etc.		NM/1 or M/1
Non-residential	Museums, art galleries, shops, offices, libraries, town halls.	Where lighting may be dimmed	M/1
		Other	M/1 or NM/1

NOTE that, as stated above, M indicates a maintained luminaire. NM indicates a non-maintained luminaire. The digit after '/' indicates the period (hours) of illumination after mains failure.

CHAPTER ELEVEN

Part P of the Building Regulations

11.1 THE BUILDING REGULATIONS

The Building Act, 1984 requires that all buildings in England and Wales must comply with the Building Regulations. The purpose of this law is to make sure that

1 the buildings are structurally sound
2 they are safe and convenient for people to use
3 they conserve fuel and power as far as is possible
4 they prevent waste, undue consumption misuse or contamination of water.

In common with the electrical regulations (BS 7671), they are written in a legalistic form that sometimes makes their requirements difficult for the practical person to understand. They consist of fourteen parts, each dealing with a different aspect of building. Designers and constructors of buildings are required by law to obey them, and a Building Control Service exists to ensure that they do so.

11.2 THE ELECTRICAL REGULATIONS

The IEE Wiring Regulations, now BS 7671, have always been a legal requirement for electrical installations in all situations with the exception of dwelling houses in England and Wales. This situation changed on January 1st, 2005, when Part P of the Building Regulations came into force. The Regulations were amended on April 6th 2006. In Scotland, the statutes have applied to all electrical installations for many decades. The amendment to Part P was 'to provide greater clarity' and was issued by the Department for Communities and Local Government which has replaced the Office of the Deputy Prime Minister.

So what does the advent of Part P mean for electrical installers in England and Wales? Unless the installer is a member of a 'competent person self-certification scheme', the proposed work must be notified to the Local Authority before commencement, and must be inspected and tested by them on completion and before being used. It seems likely that the expense and the delay in dealing with the Building Control Department of the Local Authority will mean that those who carry out a great deal of work in dwellings may wish to avoid this course of action and become members of a competent person scheme.

11.3 TO WHICH INSTALLATIONS DOES PART P APPLY?

Part P applies to electrical installations, or parts of installations, in buildings, or in parts of buildings, which are

1 dwelling houses or flats;

2 dwellings and business premises that have a common supply, such as shops and public houses with a flat above;

3 common access areas in blocks of flats such as corridors and staircases;

4 shared amenities of blocks of flats such as laundries and gymnasiums;

5 in or on land associated with the buildings – for example to fixed lighting and pond pumps in gardens and in outbuildings such as sheds, garages and greenhouses.

It does not apply to schools, care homes, hotels, boarding houses, halls of residence or hostels. These buildings are already subject to the Building Regulations and to BS 7671. If in doubt, the Local Authority Building Control Service should be consulted. Part P applies only to the work being carried out. There is no requirement for the existing electrical installation to be upgraded so that it is brought in line with the current Building Regulations or with BS 7671. However, where the work is an addition to or an alteration to an existing installation, work may be needed on it to ensure that protective measures, earthing, bonding and the adequacy of the mains supply comply with BS 7671.

Any building work carried out in connection with electrical work, such as chasing for conduits and cables, removal and replacement of floors, drilling and notching joists, drilling and making good holes, must be carried out to comply with the Building Regulations (see {4.4.1}).

The electrician must have an appreciation of how the Building Regulations in general affect the electrical installation and must be sufficiently competent to confirm that his work complies with all the applicable requirements of the Building Regulations. This is a complex subject and is dealt with in the following Sections {11.11 to 11.17}.

11.4 SELF-CERTIFICATION

Those following this route (and this must surely include all but the smallest of electrical installation firms) will need to become a member of a self-certification scheme, when no approach to the Local Authority will be necessary. Those registered will be deemed to be 'competent persons' to carry out the work, and to do so they will need to complete a registration procedure, which typically consists of

1 applying to one of the approved Registration bodies (see below) using the applicable form

2 paying the registration fee. An additional annual fee to retain registration is also payable. These fees are subject to VAT.

3 undergoing an inspection of samples of their work, their working staff and their technical staff

4 on satisfying the requirements, they will become a registered member and a 'competent person' so that they can self-certify their work.

The five bodies approved by the Department for Communities and Local Government (formerly the Office of the Deputy Prime Minister) to register competent persons self-certification schemes are

1 ELECSA, (British Board of Agremont) 44-48 Borough High Street, London, SE1 1XB. Tel: 0845 634 9043, fax 0870 749 0085, www.enquiries@elecsa.org.uk

2 BRE (Building Research Establishment) Certification Ltd. Bucknalls Lane, Garston, Watford WD25 9XX. Tel: 01923 664 0000, fax 01923 664 603, www.enquiries@bre.co.uk

3 BSI Product Services, Marylands Avenue, Hemel Hempstead, Herts, HP2 4SQ. Tel: 01442 230 442, fax 01442 278 636, www.product.services@bsi-global.com or www.kitemarktoday.com

4 NICIEC Certification Services Ltd, Warwick House, Houghton Hill Park, Houghton Regis, Dunstable, Beds, LU5 5ZX. Tel: 0870 013 0382, fax 0207 564 2370, www.enquiries@niciec.com

5 NAPIT (The National Association of Professional Inspectors and Testers), The Gardeners Lodge, Pleasley Vale Business Park, Mansfield, Notts, NG19 8PL. Tel: 0870 444 1392, fax 0870 444 1427, www.info@napit.org.uk

Electrical Installation work in dwellings that is carried out by a 'Competent Person for Part P' does not need to involve the Building Control Service. On completion of the work, the building owner must be provided with a certificate stating that the work has been carried out in compliance with the Building Regulations. The Local Authority must also be provided with a description of the work, where and when it was carried it out, and by whom. Some operators of Competent Person Schemes will process this paperwork for their members when notified.

11.5 NON-CERTIFIED ELECTRICIANS

Non-certified electricians or others, such as plumbers and kitchen fitters, can still carry out electrical work, but they will need to notify the appropriate Building Control Body (usually the Local Authority) and obtain their permission before commencing work. When complete, the work must be inspected by the control body. Fees, which are subject to VAT, will be payable for these services.

There are two procedures, one of which must be followed.

1 full plans application. Full plans and details of the proposed work are required, preferably well in advance of the proposed start date. These will be checked by the local authority, which will issue an approval within five weeks. Alternatively, a conditional approval may be issued, which will specify the modifications that must be made to the plans or will indicate what additional plans are required. If the plans are rejected, the reason will be stated. When the work is finished and inspected, a completion certificate will be issued.

2 Building Notice Procedure. This procedure is best suited to small projects, and is the type most likely to be used by the non-certified electrician. The Local Authority is given a Building Notice and work can commence at once. There is no legal requirement for a completion certificate, although in most cases one will be issued. The person carrying out the work must issue a commencement notice to the Local Authority at least two clear working days before work starts.

It has long been a heart-felt complaint by fully-qualified but not registered electricians and electrical engineers that they were prevented by the requirements of Part P from carrying out domestic electrical installation work. The 2006 amendment refers specifically to installers not registered with a Part P competent person self-certification scheme but qualified to complete BS 7671 installation certificates. It seems that such people will still need to notify the Local Authority before commencing work, who will accept their completed certification.

Unregistered installers should not themselves arrange for a third party to complete final inspection and testing. The third party, not having supervised the work from the outset, would not be in a position to verify that the installation work complied fully with BS 7671. Only the installer responsible for the work can issue an electrical installation certificate.

11.6 THE PENALTIES

Contravening the Building Regulations is a criminal offence and the person carrying out electrical work that contravenes them can be fined

1 up to £5,000 for the contravention
2 £50 for each day that the contravention continues

If work fails to comply with the Building Regulations, (including failure to notify and obtain prior permission) Local Authorities have the power to require its removal or alteration. Householders may also encounter problems in the future when offering their houses for sale. The final responsibility for ensuring the electrical work complies with the Building Regulations lies with the householder/ owner, and they may be served with an enforcement notice. However, the primary responsibility for ensuring compliance rests with the person carrying out the electrical work.

Building inspectors do not have enforcement powers, but may refuse to issue the final certificate where it is considered that there is a case of non-compliance.

11.7 DIY WORK

Do it yourself (DIY) work is still permitted after the introduction of Part P, but a building notice must be submitted to the Local Authority before work starts. On completion, the work must be inspected and tested by them, for which a fee (plus VAT) must be paid.

The number of people carrying out such work who will be prepared to go to these lengths is questionable. This suggests that illegal work is likely to be carried out unless the householder is prepared to have the work done by a competent electrician. Many electrical contractors will share the opinion of the author that it is wrong for electrical installation equipment, such as consumer units, cables, socket outlets and so on, to be on sale to the general public in stores, when their installation is illegal under Part P.

11.8 PUBLICITY

If the rules applying to Part P are to be generally obeyed, it follows that the general public must be made aware of their existence. There are plans for scheme operators, retailers, wholesalers and the Department of Communicaties and Local Government to make members of the public aware that work on fixed electrical installations in dwellings is subject to the requirements of the Building Regulations. Such publicity has been sparse.

11.9 WHAT NEED NOT BE NOTIFIED?

1 Any electrical installation work carried out by a firm or an individual registered under an authorised competent person self-certification scheme.
2 Electrical installation work that is restricted to the type described in [Table 11.1] and does NOT include the provision of a new circuit. Certain work carried out in kitchens or in special locations as shown in [Table 11.2] is excluded from [Table 11.1] and is therefore notifiable.
3 All replacement work is non-notifiable even when carried out in a kitchen or in another special location shown in [Table 11.2].

Table 11.1 Work that does NOT need notification

1   replacing accessories, switches, ceiling roses, socket-outlets, etc.
2   replacing the cable for a single circuit after damage (Note A)
3   providing mechanical protection to existing installation items (Note B)
4   replacing or refixing enclosures of existing installation items (Note C)
5   all work in caravans
6   external light or fan on dwelling wall if the switch is inside
7   replacement of cooker, shower or immersion heater of same rating
8   minor work (not a new circuit) in an attached garage
9   minor work (not a new circuit) in a conservatory
10 installing or upgrading main or supplementary equipotential bonding
11 wall lights, air conditioning units and radon fans fitted externally so long as the
    electrical connections are made directly within the enclosure of the equipment.
12 connection of electrically operated garage doors or gates to existing isolators
13 work NOT in a kitchen or in other special locations (Note D) and consisting of
    1 adding lighting points to an existing circuit (Note E)
    2 installing or upgrading main or equipotential bonding

Notes to Table 11.1

Note A   Only if the replacement cable has the same current carrying capacity, follows the
         same route and does not serve more than one circuit through a distribution board.
Note B   Only if the circuit's protective measures are not affected.
Note C   Only if added thermal insulation has no effect on current carrying capacity and on
         circuit electrical protection (fuses or circuit breakers).
Note D   See [Table 11.2] for details of locations that are special.
Note E   Only if the existing circuit protective device is suitable and provides proper
         protection for the modified circuit.

11.10 WHAT MUST BE NOTIFIED?

The work to be notified includes the following

1   all work not covered by the exemptions listed in {Table 11.1}. This will, in
    practice cover almost all of the work normally carried out by the electrical
    contractor including replacement of the consumer unit.
2   minor work carried out in the special locations listed in {Table 11.2}.

Note [Chapter 7] and IEE Guidance Note 7 give considerable help in completing
safe installations in these special locations.

Table 11. 2 Special locations and installations

1   Kitchens.
2   Locations containing a bathtub or shower basin.
3   Detached garages.
4   Electric floor or ceiling heating systems.
5   Hot air saunas.
6   Swimming and paddling pools.
7   Extra-low voltage lighting installations, other than pre-assembled CE marked lighting
    sets.
8   Solar voltaic (PV) power supply systems. (see section {8.22})
9   Small scale generators such as micro CHP units. (see section {8.21})
10 Garden lighting and power supplies.
11 New central heating control wiring regardless of the location of the cabling.
12 Replacement of the consumers unit.

11.11 BUILDING REGULATIONS APPROVED DOCUMENT A – STRUCTURE

The important message here is that alterations to the building construction carried out by the electrician during installation work must not result in dangerous weakening of the building structure. There are two matters here to which the electrician must pay attention.

1 Notches and holes in floor and roof joists.
2 Chases in structural walls.

The requirements for notches and holes in joists are covered in section {4.4.1}, as are those for chases in walls. Notches or holes must not be cut in roof rafters.

11.12 BUILDING REGULATIONS APPROVED DOCUMENT B – FIRE SAFETY

Detailed information on the lighting of escape routes can be found in {Chapter 10} Emergency Lighting.

This part of the Building Regulations is very concerned with the ability of the structure to contain fire and to prevent its spread. In this respect, it is very important that there should be no holes in the structure through which fire and smoke can propagate. Electricians have sometimes been culprits in this respect by cutting oversized holes for the passage of cables and their enclosures and then in failing to seal them up afterwards. For guidance on correct practice, see Sections {4.4.1} and {4.5.2}.

Smoke alarms are also a requirement of the Building Regulations. For information on this subject, see Section {2.2.4}. Fire alarms are becoming more common in dwellings, although not yet a requirement of the Building Regulations. They are considered in some detail in {Chapter 9} Alarm Systems.

Where it is critical for electrical circuits to continue to function in the event of fire, protected circuits are needed. Such circuits must be

1 wired in special fire-resistant cables to BS 6387
2 routed to pass through only parts of the building with negligible fire risk
3 separated physically from all other circuit cables.

11.13 BUILDING REGULATIONS APPROVED DOCUMENT E – RESISTANCE TO THE PASSAGE OF SOUND

The Building Regulations require that a dwelling must be so constructed that there is an adequate means of protection against sound, both generated internally and externally. They also require protection from the reverberations, that often result from sound of high intensity but low frequency. The insulation used for this purpose within ceilings and other situations in which cables are run often also has thermal insulation properties, and will sometimes have the effect of reducing the current-carrying capacity of such cables, as explained in Section {4.3.6}.

Recessed luminaires installed in ceilings will often have a deleterious effect on sound insulation. It follows that electrical work must never adversely affect the sound insulation properties of a building.

11.14 BUILDING REGULATIONS APPROVED DOCUMENT F – VENTILATION

The only rooms where ventilation is a requirement of the Building Regulations are kitchens, utility rooms and bathrooms. Table 1 of the Document F applies to rooms with openable windows that are located on external walls. In the case of rooms without openable windows, the same figures apply, but fans must have a 15-minute overrun. Such rooms are likely to need artificial light and it is usual for the fan to be started by the light switch. Some extract rates for common domestic situations are given in {Table 11.3}.

In practice, the electrician is unlikely to be able to measure the extract ventilation rate, his only method of control being a sensible choice of fan size. Where combustion appliances (such as gas boilers) are in use, ventilation may be necessary to prevent the build-up of dangerous gases such as carbon monoxide. The combustion appliance itself usually meets the requirement in these cases, but the electrician should bear in mind the need for care.

Table 11.3 Extract ventilation rates

Room	Extract rate (litres/second)
Kitchen – adjacent to the hob	30
Kitchen – elsewhere	60
Utility room	30
Bathroom (with or without WC)	15

11.15 BUILDING REGULATIONS APPROVED DOCUMENT L1 – CONSERVATION OF FUEL AND POWER

It has been claimed that the emission of 'greenhouse gases' by electrical power stations is a major contributor to global warming. The Building Regulations thus require that electrical energy is used as efficiently as possible. These energy conservation requirements apply to:
1 all new dwellings
2 extensions to existing dwellings
3 loft conversions
4 new outdoor lighting installations.

Internal lighting
A method of meeting the requirements is to provide a proportion of the fixed lighting with fluorescent or compact fluorescent lamps, which have a higher efficiency than general lighting service (filament) lamps. At least one third of luminaires or pendants should be of this type, with a luminous efficiency of 40 lumens per watt or higher. Such outlets must not be provided with BC or ES lampholders, to prevent replacement of high efficiency lamps with less efficient types.

External lighting
There are two suggestions for reducing the energy used by external lighting fixed to dwellings (including porches but not garages or car ports). They are the
1 provision of photo-cell control to prevent their use in daylight
2 use of high-efficiency lamps such as fluorescent or compact fluorescent types.

11.16 BUILDING REGULATIONS APPROVED DOCUMENT M – ACCESS TO AND USE OF BUILDINGS

These Regulations apply to new buildings and are concerned with their accessibility and their use. The electrician can have little control of the former, but the positions of switches and socket outlets can have a considerable effect on building use, particularly by those who are disabled such as wheelchair users. The requirements will be met by ensuring that all switches and socket outlets are positioned between 450 mm and 1200 mm from floor level.

11.17 BUILDING REGULATIONS APPROVED DOCUMENT P – ELECTRICAL SAFETY

Part P applies to all electrical work carried out in new or existing fixed electrical installations operating at low voltage or at extra-low voltage in dwellings and associated areas. This Chapter has been concerned with the effects of Part P on electrical installation work, and attention is drawn particularly to sections {11.9} and {11.10} dealing with minor works and special locations.

11.18 DWELLINGS SUBJECT TO FLOODING

Although not included in Part P of the Building Regulations, electricians need to take account of the possibility of flooding in houses they are wiring. The prevalence of flooding seems to be more common and may well increase if global warming becomes more pronounced.

Some items to bear in mind when installing or altering electrical wiring in such situations are:

1 Cables supplying lower floor power should be routed through an upper floor to reduce the effect on them in the event of a flood

2 All electrical equipment should, as far as possible, be mounted above the expected flood level.

3 Cables likely to be under water in the event of flood should be drawn into plastic conduit. The provision of these conduits may reduce the amount of rewiring necessary after a flood. Drain holes must be provided at the lowest levels of such conduits to allow drainage to reduce the risk of long-term water damage to the cables.

Regulation 24 of the Electrical Safety, Quality and Continuity Regulations (ESQCR) 2002 requires Electricity Distributors to install and, so far as is reasonably practicable, their equipment to prevent danger, which can include the risk of flooding.

List of abbreviations

A	ampere – unit of electric current	I_a	current to operate protective device
BC	bayonet cap		
Ca	ambient temperature correction factor	I_b	design current
		I_d	fault current
CENELEC	European Committee for Electrotechnical Standardisation	IEC	International Electrotechnical Commission
C_g	cable grouping correction factor	IEE	Institution of Electrical Engineers
C_i	thermal insulation correction factor		
C_t	correction factor for the conductor operating temperature	I_n	current setting of protective device
		I_t	tabulated current
CNE	combined neutral and earth	I_z	current carrying capacity
cos ø	power factor (sinusoidal systems)	IT	earthing system (see 5.2.6)
		k	kilo – one thousand times
CPC	circuit protective conductor	kV	kilovolt (1000 V)
c.s.a.	cross-sectional area	L1, L2, L3	lines of a three-phase system
D_e	overall cable diameter	m	metre
ECA	Electrical Contractors Association	m	milli – one thousandth part of
		M	meg or mega – one million times
EEBAD	earthed equipotential bonding & automatic disconnection	mA	milliampere
ELCB	earth leakage circuit breaker	MCB	miniature circuit breaker
ELV	extra-low voltage	MCCB	moulded case circuit breaker
EMC	electro-magnetic compatibility	MD	maximum demand
e.m.f.	electro-motive force	m.i.	mineral-insulated
EMI	electro-magnetic interference	NICEIC	National Inspection Council for Electrical Installation Contracting
ES	edison screw		
f	frequency	p.d.	potential difference
FELV	functional extra-low voltage	PELV	protective extra-low voltage
GN	guidance note	PEN	combined protective and neutral
HBC	high breaking capacity (fuse)	PME	protective multiple earthing
HRC	high rupturing capacity (fuse)	PIR	passive infra-red detector
Hz	Hertz – unit of frequency	PSC	prospective short-circuit current
I	symbol for electric current	p.v.c.	poly-vinyl chloride
I_2	operating current (fuse or circuit breaker)	R	resistance (electrical)

R	resistance of supplementary bonding conductor	U	symbol for voltage (alternative for V)
R_A	the total resistance of the earth electrode and the protective conductor connecting it to exposed conductive parts	Uac	alternating voltage
		Udc	direct voltage
		Uo	phase voltage
R_p	resistance of the human body	V	volt – unit of e.m.f. or p.d.
RCCB	residual current circuit breaker	W	watt – unit of power
RCD	residual current device	X	reactance
r.m.s.	root-mean-square (effective value)	Z	impedance (electrical)
		Z_e	earth loop impedance external to installation
s	second – unit of time		
S	conductor cross-sectional area	Z_s	earth fault loop impedance
SELV	safety extra-low voltage		thermal conductivity
t	time	ø	phase angle
TN-C	earthing system (see 5.2.5)	Ω	ohm – unit of resistance, reactance and impedance
TN-C-S	earthing system (see 5.2.4)		
TN-S	earthing system (see 5.2.3)	µ	micro – one millionth part
TT	earthing system (see 5.2.2)		

Index

BP 2.09   BC 3/11